Meredith Clymer

Notes on the physiology and pathology of the nervous system

With reference to clinical medicine

Meredith Clymer

Notes on the physiology and pathology of the nervous system
With reference to clinical medicine

ISBN/EAN: 9783742830494

Manufactured in Europe, USA, Canada, Australia, Japa

Cover: Foto ©Klaus-Uwe Gerhardt /pixelio.de

Manufactured and distributed by brebook publishing software
(www.brebook.com)

Meredith Clymer

Notes on the physiology and pathology of the nervous system

NEW YORK:
D. APPLETON & COMPANY,
90, 92 & 94 GRAND STREET.
1870.

NOTES

ON

THE PHYSIOLOGY AND PATHOLOGY

OF THE

NERVOUS SYSTEM,

WITH REFERENCE TO

CLINICAL MEDICINE.

INTRODUCTION.

It is proposed in these papers to summarize the recent investigations into the physiology and pathology of the nervous system which have a bearing on clinical medicine. Quite lately much of worth has been done in this direction. New and nicer methods of examining the intimate structure of the brain and spinal marrow in health and in disease, have unsettled many traditional notions concerning several classical disorders, and fixed the connection between symptoms and lesions. Thanks to the better studies of those who have busied themselves with these subjects, we have now satisfying and trustworthy information, where before confusion and uncertainty, or, at best, happy conjecture, ruled. There has been patient and productive working; facts have been largely and heedfully gathered, and rightly used; sagacity has tempered zeal; and more concern has been shown about practical results than engaging hypotheses. It is not, therefore, claiming too much to say that art is a great gainer by these scientific inquiries. The outcome is a less faulty pathogeny, with a truer understanding of many common affections of the nervous centres, and surer means of detection and treatment.

I.

SCLEROSIS OF THE NERVOUS CENTRES.

Virchow says, "Sclerosis (σκληρός) signifies thickening with condensation."[1] This definition is too limited and vague. The term is strictly descriptive, not of induration from any cause, but of changes in the textural condition of an organ. It involves both quantity and quality. There is overgrowth (proliferation) and transformation of connective tissue, with consequent wasting of the proper functional elements of the part. The processus is constant wherever it happens, and includes both creation, metamorphosis, and destruction. The undue development of the basis-tissue in the liver, kidney, or lung, is always at the cost of the specific material of those organs; so sclerosis of the nervous centres means not only parasitic exuberance of the connective gangue (neuroglia), but proportional compression, deterioration, and annihilation of the ganglion-cells and nerve-tubes. There is histological substitution. This is its univocal characteristic. The perversions of function which we shall study under the name of symptoms are all referrible to it. Comprising what has hitherto been treated of by writers under the name of chronic inflammation of the brain and spinal cord—chronic encephalitis and chronic myelitis,—it is only lately that its true pathogeny has been intelligently set forth. Morbid conditions, hitherto confounded but essentially distinct, have been differentiated, and the proper signs of each modality established.

Sclerosis of the nervous centres is met with in four forms: *A.* Disseminated, or Multilocular; la sclérose en plaques disseminées; *B.* Cortical, annular, or peripheral; la sclérose ou annulaire. *C.* Fascicular, *a*, primitive, *b*, secondary; la sclérose rubannée, primitive ou secondaire; *D.* Diffuse; la sclérose diffuse.

[1] Cellular Pathology, Chance's translation. American Reprint, p. 468.

A. DISSEMINATED, OR MULTILOCULAR, SCLEROSIS OF THE BRAIN AND SPINAL CORD.

Definition.—*A disease of the cerebro-spinal centres; of gradual invasion, beginning with muscular weakness of one or both lower limbs, subsequently extending to the upper, and, sooner or later, passing into complete paralysis, which may in time affect, in some degree, the muscles of the head, neck, face, pharynx, and tongue; no constant derangement of cutaneous sensibility; the univocal symptom, tremor in the implicated muscles, which happens only when any intentional movement is attempted, and ceases in a state of rest; frequently nystagmus; attended in the later stages with cramps, and permanent rigidity and contraction of the palsied members; of probably diathetic origin; slowly and surely progressive in its course, and constantly of fatal termination; the anatomical characters being patches or corns of sclerosis, irregularly disseminated, in the brain and spinal cord.*

History.—The first mention of a pathological condition of the nervous centres which was probably sclerosis, is in a report of "*A Palsy occasioned by a Fall, attended with Uncommon Symptoms,*" by Dr. A. Maty, dated December 29, 1766, and published in "*Medical Observations and Inquiries,*" vol. iii., 2d edition, London, 1769, p. 257. Count Lordat, a French nobleman, had, in the overturning of a carriage, his neck sharply twisted. He quickly apparently recovered from the effects of the accident, and went through a severe military campaign, during which he was constantly exposed to damp and wet. Some time after he noticed an impediment in his speech; then his left arm grew weak, and the corresponding leg; he began to have convulsions all over the body, and became paralytic. The account of the autopsy was sent to Dr. Maty by the count's physician in France. He writes: "We observed no signs of compression in the lingual and brachial nerves as high as their exit from the basis of the cranium and the vertebræ of the neck; but they appeared to us more *compact than they commonly are, being nearly tendinous.* . . . The central substance *of the brain appeared much browner than usual.* . . . We chiefly took notice of the medulla oblongata; it was *more compact . . . the marrow itself had acquired such solidity as to elude the pressure of our fingers;* it resisted as a callous body, and could not be bruised. This hardness was observed all along the vertebræ of the neck, but lessened by degrees, and was not near so considerable in the vertebræ of the thorax." By the term vertebræ, the region of the spinal cord is evidently meant.

Dr. Marshall Hall (*Diseases and Derangements of the Nervous System*, 1841) relates the case of a man, aged twenty-eight, with tremor of the right arm and leg, who had a peculiar rocking motion of the eyes, and a degree of stammering and defective articulation. This was unquestionably an example of Charcot's disease.

The earliest description of disseminated cerebro-spinal sclerosis is by Cruveilhier, in his *Atlas de l'Anatomie Pathologique* (Paris, 1835–1842). Two cases are given; and their clinical history and morbid anatomy, with the accompanying plates, leave no doubt of their nature.[2] About the same time (1838) Carswell gave an accurate representation of it in his *Illustrations of the Elementary Forms of Disease*, under the head of "Atrophy" (pl. iv., fig. 4). In the descriptive note he says: "Isolated points of the pons varolii, of a yellowish-brown color, patches of the same kind on the spinal chord; all of them occupying the medullary substance, which was very hard, semi-transparent, and atrophied. The atrophy was more conspicuous in some points than in others." In 1856, Ludwig Türck published several cases.[3] Rokitansky

[2] Obs. Paraplégie: Degénération grise de la moelle, du bulbe, de la protubérance, des pédoncules cérébelleux, des couches optiques, du corps calleux, de la voûte à trois piliers.—Liv. xxxii., pl. ii., fig. 4.

Obs. Insensibilité presque complète des membres inférieurs. Diminution de la sensibilité des membres supérieurs. Diminution notable, mais relativement moindre, de la myotilité.—Dégénération grise beaucoup plus considérable dans les cordons postérieurs que dans les cordons antérieurs de la moelle épinière.—Liv. xxxviii., pl. i. and ii. A mixed case, being associated with posterior sclerosis, or locomotor ataxy, and to be more particularly mentioned hereafter.

[3] Türck L. Ueber Degeneration einzelner Rückenmarksstränge. Sitzungsbericht der K. K. Akad. zu Wien; Mathem. Naturw. Classe. 1856.

mentions the structural change in his Pathological Anatomy.[4] In 1855, Frerichs (Häser's Archiv, Bd. x.) published several cases and described the lesion. The following year (1856), Valentiner, his assistant, made the first attempt at a systematic clinical history of the disorder.[5] He reprinted Frerichs's cases, and collected a number of others (15), which he supposed to be of the same kind. But, according to Dr. Charcot, he did not group them with any exactness, and, under the same title, described conditions quite dissimilar. Hasse, in Virchow's *Handbuch*, and Niemeyer in *Lehrbuch der Special. Patholog. und Therap.*, have followed Valentiner's mixed description. Jaccoud says that Demme published a case of sclerosis of the antero-lateral columns, with microscopical examination, in 1859, but where he does not state, and I have been unable to find it. Prof. Oppolzer, in 1861,[6] published a case of paralysis agitans, with induration of the pons, medulla oblongata, and of the lateral columns, particularly in the lumbar region. In Leyden's well-known contribution on gray degeneration of the cord (1863), the case numbered XXXI. is one of diffuse sclerosis.[7] Frommann reproduced this case in his work on the normal and pathological anatomy of the spinal cord, and details with great minuteness the histology of the lesion.[8] Rindfleisch, too, has described its histology with exactness,[9] and

In previous papers (1851, 1853, and 1855), he had described secondary sclerotic degenerations of the cord. In this he treats of primary sclerosis of the lateral and antero-lateral columns, and gives three cases, with microscopic examinations.

[4] Rokitansky, C. Lehrbuch der Pathol. Anat., Zweiter Band, p. 488, 1856. Ueber Bindegewebs-Wucherung im Nervensystems, Wien, 1857.

[5] Valentiner. Ueber die Sclerose des Gehirns and Rückenmarks. Deutsche Klinik, Bd. xiv., 1856.

[6] Oppolzer. Wiener medicin. Wochenschrift, 36, 38, 1861. Canstatt's Jahrb., iii., 78, 1861. Trousseau's Clinical Medicine, Bazire's translation, vol. i., 446.

[7] Die graue Degeneration der hinteren Rückenmarksstränge, Berlin, 1863. Also Virchow's Archiv. xxix., 202, 1864.

[8] Frommann. Untersuchungen über die normale u. patholog. Anatomie des Rückenmarks, 2 Theil. 77, pl. i., Jena, 1867.

[9] Rindfleisch. Histolog. Detail zu der grauen Degeneration von Hirn u. Rückenmarks. Virchow's Archiv., Bd. xxvi., 1863.

Zenker published (1865) a very carefully-drawn-up case.[10] A case of Skoda (1862) should have been also mentioned.[11] L. Leo's valuable cases were published in 1868.[12]

In France, Drs. Charcot and Vulpian had studied its anatomical characters between 1863 and 1865. On the 25th January, 1865, Charcot read a paper before the Société Médicale des Hôpitaux de Paris, on "Sclerosis of the Lateral Columns of the Spinal Cord, in an Hysterical Woman, whose four limbs had, during life, been permanently contracted." It appeared in the *Gazette Hebdomadaire*, February 17, 1865. The first complete and accurate description of the disease, clinically and anatomically, was, however, by Dr. Vulpian, in 1866.[13] Three cases were given, one of his own, and two of Dr. Charcot's. Dr. C. Bouchard utilized Charcot's and Vulpian's cases, which he had seen at La Salpêtrière, and made a communication on the subject to the Medical Congress at its meeting at Lyons, 1867. In 1866, Dr. Jaccoud gave a clinical lecture at the Hôpital de la Charité on a a case which he styled diffuse sclerosis, and which has been since published.[14] In 1868, Dr. Ordenstein's thesis appeared, in which Charcot's views were represented, and the differential diagnosis between paralysis agitans and disseminated sclerosis was for the first time[15] set forth. In the spring and summer of 1868, Dr. Charcot largely and minutely discussed the subject in his Clinical Lectures at La Salpêtrière. These were subsequently published.[16] A case was communicated to the Société de Biologie, in January, 1869, by Joffroy. The case

[10] Zenker. Ein Beitrag zur Sclerose des Hirns u. Rückenmarks, Zeitschr. f. rat. Medizin. Bd. xxiv.

[11] Skoda. Wien. Med. Halle, iii., 13, 1862. Schmidt's Jahrbucheck, No. 119, p. 294. Syd. Soc. Year Book, 1863, 100.

[12] Leo. L. Deuts., Archiv. f. Klin. Med., 1868, 151.

[13] Vulpian. Note sur la Sclérose en plaques de la moelle épinière. Union Médicale, 1866.

[14] Jaccoud. Leçons de Clinique Médicale, 2d ed., Paris, 1869. In Les Paraplégies et l'Ataxie du Mouvement (p. 245), 1864, he had already alluded to the subject. See also Traité de Pathologie Interne, t. i., 321, Paris, 1869, by the same author.

[15] Ordenstein. Sur la Paralysie agitante et la sclérose en plaques généralisées. Paris, 1867.

[16] Gazette des Hôpitaux, 1868.

of Dr. Pennock, by Drs. Morris and Weir Mitchell, appeared in the *American Journal of the Medical Sciences*, July, 1868. It is the only contribution of this country. It does not appear that the diagnosis was made during life. In 1869, Drs. Bourneville and Guérard published their essay, "De la Sclérose en Plaques Disséminées;" and subsequently Dr. Bourneville his "Nouvelle Étude sur quelques Points de la Sclérose en Plaques Disséminées." They are a reproduction of Charcot's Clinical Lectures, with a collection of all the known cases of the affection. In the last volume of the Memoirs of the Biological Society of Paris, Dr. Liouville has two interesting cases, and Dr. Magnan one, in which there was papillary atrophy of both eyes.[17]

To Dr. Charcot, therefore, unquestionably belongs the credit of distinguishing this affection from other paralytic disorders, and notably from paralysis agitans, of recognizing its pathological individuality, and tracing its clinical history. He has done for it what Chomel and Louis did for typhoid fever when they established it as a distinct species of continued fever, characterized by a definite group of symptoms.

There has always been much confusion in the clinical history of the disorders of the nervous system in which tremor is a chief symptom; and the result has been to confound diseases pathogenetically distinct. A proper comparative study of the different forms of tremor, and an accurate investigation of the quality of the symptom have been only recently made. There are two morbid groups in which tremor is a prominent symptom, and in each its character is proper and distinct. In the one it is a constant symptom, or, if temporarily suspended, it is only during sleep; in the other it never happens except as accompanying a voluntary movement; it is always absent when the limb or body is at rest. Galen seems to have noted the two kinds, and to have made a distinction between τρόμος (tremor) and παλμὸς (tremor coactus). Van Swieten speaks of tremor coactus, always present except during sleep, and tremor debilitate, which accompanies intentional movements. (Comment. Aph., 625.) Dr. Gubler recognized also the two varieties of trembling, but made no attempt at any clinical or pathological interpretation (*Archives Gén. de Méd.*, 5ᵉ s., t. xv., p. 702, 1860).

[17] Liouville. Deux cas de sclérose en îlots multiples et disséminés du cerveau et de la moelle épinière. Comptes rendus des séances et Mémoires de la Société de Biologie, t. xx., 1869.

Magnan. Note sur une observation de sclérose en plaques cérébro-spinale avec atrophie papillaire des deux yeux. l. c. Archives de Phys. Nor. et Path. 2ième année (1869) p. 765.

All the English authors confound this disorder with paralysis agitans. Parkinson, whose description of shaking palsy has been closely followed, unquestionably did. He was aware of Maty's case, refers to it, and expresses the opinion that the probable morbid condition in paralysis agitans may prove to be induration of the upper part of the medulla spinalis, oblongata, and pons, due, as he supposes, to simple inflammation, or rheumatic, or scrofulous affection of the nervous substance or membranes. Dr. W. R. Saunders (Reynolds's "System of Medicine," vol. ii.), in an excellent article on paralysis agitans, confuses it with diffuse cerebro-spinal sclerosis; though aware of the sclerotic lesion of the cord, he is greatly at fault when he attempts to connect it with its proper clinical history. For example, he says: "In more inveterate, especially senile cases, paralysis agitans appears to depend on a discoverable lesion, namely, an atrophic condition of the spinal cord," etc. (p. 199). Again: "This atrophy would explain the chief features of the disease. . . . the occurrence in old age, . . . under conditions of premature senility" (p. 199). Previously (p. 186) he writes: "Lastly, the term paralysis agitans, or shaking palsy, has been applied to cases of ordinary motor paralysis (hemi- and paraplegia) complicated with tremors—a complication not uncommon in diseases of the brain, and in certain cases of chronic myelitis, and of locomotor ataxia. . . . Parkinson's malady is *idiopathic* paralysis agitans, in which the tremors are the chief and earliest symptom, and the paralysis entirely subordinate and peculiar, true hemi- or paraplegia being rare complications; while, in the cerebral and spinal affections just referred to, the loss of motion or sensation is the main feature of the disease, and the tremors and spasmodic agitations are only concomitants (i. e., the paralysis agitans is *symptomatic*). Hence the latter kind of cases should be styled, not paralysis agitans, but hemi- or paraplegia, or spinal or cerebral disease *complicated with paralysis agitans;* i. e., with spasmodic tremors. This description, which is essential for the accurate definition of Parkinson's disease, has often been overlooked, and requires, therefore, to be specially insisted on." Really, though imperfectly, describing diffuse sclerosis of the nervous centres, at p. 193, he says: "Occurring in middle life (twenty-five to fifty), however formidable in appearance, it is susceptible of amelioration, and sometimes of cure." There is other evidence in the article to show how unsatisfactory the writer's notions are respecting the disease we are about to describe.

The latest British authority, Dr. Handfield Jones (Studies on Functional Nervous Disorders, 2d edition, London, 1870), treating of paralysis agitans, remarks: "It appears to me a question whether two distinct affections are not often comprehended under this name" (p. 383). But, from what immediately follows, it is clear that he had not disseminated cerebro-spinal sclerosis in mind, as one of the two indiscriminated affections. He goes on to say: "For, on the one hand, it appears pretty certain that there is one form which is *met with in old persons, is quite incurable*, and is *asso-*

ciated with, if not *dependent on*, *organic wasting changes in the nervous centres;* while another form occurs in *younger persons*, is more curable, and is therefore presumably not dependent on organic change."

Clinical History.—Although the anatomical characters of this affection are constant, and always of the same histological constitution, their territorial distribution may vary. This capriciousness of site consequently qualifies the symptoms, and it is necessary to admit several forms of the disorder, each to a certain extent represented by proper functional disturbances; these are determined by the exclusive, or predominant, occupation by the lesion of one or more districts of the cerebro-spinal system.

Disseminated sclerosis of the nervous centres may be described, therefore, under three divisions, according to situation: *a*, cerebral form; *b*, spinal form; *c*, cerebro-spinal form.

a. Cerebral Form.—It is doubtful if this form ever happens strictly alone. The only recorded observation is that by Valentiner and Frerichs, and here the spinal cord was not examined after death with proper care.

CASE.—A youth, nineteen years of age, was suddenly seized, without apparent cause, with unilateral (left) motor and sensory paralysis, which soon extended to the other side. There was general muscular unsteadiness, and voluntary movements brought on excessive tremor in the extremities, which after a while was induced by any attempt at speaking or by moral emotion. The gait was oscillating, and the patient obliged to use crutches. At first, some mental excitement was noticed, but there soon succeeded spells of melancholy. There were eccentric neuralgia and twitchings in the affected muscles, and subsequently attacks of vertigo, with pain in the occipital region. Two years after the outset, paralysis of the limbs was almost complete, though still most marked on the left side; the tremor had increased and was nearly constant; speech unintelligible; urination and defecation irregular; nystagmus; and gradual weakening of the intellect. A few days before death, which occurred about this time, tremor ceased. Patches of sclerosis were found in the pons varolii, olivary bodies, and at the base of the brain.

b. Spinal Form.—The invasion is usually very gradual, the first symptoms being tingling and numbness in one or both legs, or soles of the feet, which are soon followed by weakness in the limbs; this last may, however, be an initial symptom. The paresis grows worse by degrees, and if it has been limited to one extremity, as is often the case, will, after a while, involve the other, and extend to the upper extremities. The gait, which from the outset may have been more or less unsteady, is now staggering like that of a drunken man. All muscular acts are uncertain, and rhythmical spasms accompany any voluntary movement of the affected muscles. Cutaneous sensibility, in respect to touch, pain, and temperature, is unaffected, the cases in which, at this time, any modification has been noticed, being exceptional. The general health is good. As the disease advances, paralysis, more or less complete, succeeds the paresis; all the symptoms just described worsen, and new ones appear. They are: 1. Tonic muscular spasms, occurring spontaneously, or after artificial excitation. Although generally happening at a late period, they have been met with during the early stages, particularly in those cases in which paraplegia has set in soon. These cramps are mostly limited to the lower extremities, rarely affecting the upper, and, when they do, only in a slight degree. One or both lower limbs may become suddenly rigid, and any attempt to move them in certain directions is found to be difficult or impossible. After some time this rigidity passes off, and the affected member may be moved at the will of the operator. 2. Permanent contractions, which follow strictly in their development the course of the palsy, striking first the legs, then the arms, and occasionally the muscles of the trunk. The position of the lower extremities is, for a while, that of fixed extension, while the fingers are bent inwards. Finally, motility, which has been progressively growing weaker, is totally abolished; all power of voluntary motion in both the upper and lower extremities is lost, and the unfortunate patient is condemned to keep his bed. The legs are forcibly flexed on the thighs, the thighs on the pelvis, and the heels are drawn closely up towards the buttocks. It is almost impossible to straighten the limbs, an effort to do so causing great pain. Sensibility is still often intact. In some patients, reflex movements may be provoked by pinching the skin, or tickling the soles of the feet, while in others these excitations are without result. Finally, the general health fails; nutrition becomes defective, and there is rapid emaciation; sloughs form on the sacrum, and death happens from exhaustion, or some intercurrent acute disease.

There are several varieties of this form of spinal sclerosis, as one or other column of the cord is chiefly or exclusively affected; or more than one column may be implicated, and in such cases the symptoms will be of a mixed character, as when the lesion involves the posterior and anterior, or antero-lateral, columns at the same time. In such cases the phenomena proper to locomotor-ataxy will be associated with those of the spinal form of disseminated sclerosis. (See Cases in Appendix.)

c. Cerebro-spinal Form.—The accession may be insidious, as in the preceding form, or abrupt. Simultaneously with, or foregoing, or following, the motorial troubles, there are ocular or cerebral disorders, as constant vertigo, pain in the head, difficult articulation, weakness of sight, or diplopy; these may be only transitory. In some instances there have been occasional attacks of cerebral congestion, without loss of consciousness, but attended by temporary hemiplegia (Charcot, L. Leo). Sooner or later tremor affects the extremities and eyeball (nystagmus). The paralytic symptoms follow very much the same course as in the spinal form, attended with rigidity and permanent contraction. These latter have appeared as early as two years after invasion, but most generally not sooner than five or six years, so that if the patient should die from an acute disease before that time they may be altogether wanting. The spasmodic jerkings of the paralyzed muscles, spontaneous or induced, are nearly always limited to the inferior extremities. The ocular troubles get worse, and the ophthalmoscope shows slight dilatation of the retinal veins, and sometimes atrophy of the papilla. Articulation is more difficult and fragmentary; the words are as it were scanned, each syllable being distinctly and slowly pronounced, owing to muscular weakness hindering the movements necessary to proper utterance, and not to any cerebral defect. The several forms of cutaneous sensibility in many cases are undisturbed, or nearly so. The expression of the face is natural. The mental faculties may be for some time perfect, but after a while memory is impaired, the temper becomes irritable or melancholic, and intelligence grows feebler by degrees, until it is quite lost. The general functions are yet good, though there may be constipation and frequent micturition. The progressive loss of motility renders the patient perfectly helpless; the muscles of the mouth and pharynx are paralyzed, and mastication and the deglutition of solids, and even of liquids, difficult or impossible; the saliva accumulates in the mouth, or dribbles out of the corners of the lips, after ineffectual attempts at swallowing it. Sloughs form on the parts of the body exposed to pressure. Rapid deterioration of the system causes death from exhaustion; or an attack of caseous pneumonia, bronchitis, erysipelas, dysentery, or acute phthisis, or apoplexy, proves quickly fatal.

Consideration of the Special Symptoms.—*Invasion.*—In a few cases, as has been stated, the onset is without warning or apparent cause, the patient becoming suddenly paraplegic—in one instance with lessened sensibility. Or the attack may begin with an apoplectic seizure with or without loss of consciousness, and followed by temporary hemiplegia, and fits of vertigo. But in a large majority the approach is insidious, and before any weakness in the lower extremities is complained of, there is tingling in the legs and soles of the feet, with occasional numbness and coldness, and a sense of fatigue after slight exertion. Some stiffness and awkwardness in the movements may be noticed; and in a few instances, probably where the posterior columns of the cord were affected from the outset, there have been darting pain-spells.

Motility.—The paresis may be regularly progressive from the outset, or it may abate from time to time. After a while the gait becomes uncertain; the patient stumbles on level ground, or on meeting with the slightest obstacle; he gets weary quickly, and is obliged to use a cane, or some support. The loss of voluntary power in the muscles rapidly increases, and the limbs are useless. The paralysis follows invariably the course of the paresis; and frequently in this order: first, the left leg, then the right, the left upper extremity, next the right; afterward the muscles of the face, neck, and trunk.

Tremor.—This is the univocal pathognomonic symptom, to which all others are secondary, and by which the diagnosis is made. It gives the disease its special physiognomy, like the peculiar gait of locomotor ataxy. The period of its appearance is variable, and often difficult to fix. In one case it was evident three months after the onset; in one, eight months; in one, fifteen months; and in others it was not noticed until several years after the paraplegia had set in. It never occurs except on some voluntary muscular effort, or, as happens in the later stages, when excited by mental emotion, as the sight of a stranger, or the examination of the physician. At first it is slight; and, if the patient keeps his bed, may be overlooked, for it is always absent when the muscles are at rest. Under these circumstances it must be sought for; the patient should be made to execute certain movements, as carrying his hand to his head, or a vessel to the mouth. In the erect position a series of oscillations takes place, the equilibrium being maintained with difficulty, and the body swaying towards all sides. If walking is possible, titubation or staggering, as in the gait of a drunken man, is well marked, and there is a constant tendency to fall over, or sometimes back ward. Like the paralysis, tremor extends upward, and after the arms and hands, the head, eyes, and tongue, may be affected. An attempt to drink will set the head shaking, or even raising it from the pillow. The oscillatory movements of the eyeball (nystagmus) are often so great as to hinder ophthalmoscopic exploration. They usually occur only on using the eye, and trying to fix an object. The nystagmus is nearly always binocular, only one case being reported where it was monocular. When the tongue is protruded it is found to be tremulous; words are uttered in a peculiar manner, the difficulty of articulation increases, and, finally, speech is unintelligible.

The kind or type of the tremor, is a series of rhythmical twitchings (secousses) or short spasms, which at first are moderate, and do not prevent the usual movements of the affected limb, but modify their mechanism. In raising a glass of water to the lips, the gesture is not continuous and harmonious, but broken, jerking, and ill-regulated; the control of the will over the muscular effort is manifestly weakened; and the hand is carried in various directions before the act is accomplished; or, from the violence of the jactitations, it may become impossible. The tremors of the limbs are usually in the direction of flexion and extension, sometimes of ab- and adduction, and occasionally in that of rotation. The head nods, rolls, or partially rotates; and the motions of the jaw are lateral.

Towards the end, when the limbs are permanently contracted, and consequently immovable, the tremors cease in them, but may continue in the head, neck, and trunk.

Sensibility.—Sensibility to touch, pain, tickling, and temperature, may be not at all or but little deranged. Partial hyperæsthesia was noticed in a few instances. In two cases there was muscular anæsthesia, the notion of position of the limbs being lost, but cutaneous sensibility was natural (L. Leo). One patient had analgesia (Charcot); and in another contactile discrimination was lessened or wanting in certain areas of the skin (Liouville). A superficial burning heat in the lower extremities was constantly felt by one individual; and Dr. Pennock had the sensation of a narrow band around one of his legs.

Of the special senses, vision is the one most commonly affected. The sight is frequently weak from the outset; photopsia has occurred with amblyopia. Once it is stated that the sense of smell was lost, and at the autopsy patches of sclerosis were found in the olfactory nerves.

Contractions.—These never occur before the paralysis is fully established, usually after several years, and hence belong to a late period of the disease. They too follow the line of attack of the palsy; beginning in the lower extremities, they are at first partial and intermittent, but finally become generalized and permanent, often extending to the muscles of the jaws and trunk, in which case the patient lies a helpless, inert mass, drawn up in a heap. The direct relation between symptom and lesion has several times been demonstrated. In one case of contraction of the left arm, examination of the cord after death showed that the two anterior columns, and the left lateral column, at the level of the middle of the cervical enlargement, had undergone sclerotic transformation (Charcot). In another, where there had been rigidity of the right upper extremity with slight flexion of the forearm and permanent extension of the extremities, a band of sclerosis on the right antero-lateral column extended from just below the olivary body to the upper part of the cervical enlargement.

Cramps.—These occur in paroxysms, lasting from a few minutes to several days. A limb may of a sudden become thoroughly rigid, and resist all efforts to move it. One instance is mentioned where the lower extremities were seized when in a state of adduction. They are often very painful, and when over leave a sense of fatigue in the limb.

Muscular spasm-spells—the spinal epilepsy of Brown-Séquard—are almost limited to the lower extremities. The limbs are jerked about as by rapid electric shocks. When their degree is less, the spasms are like the muscular startings from strychnine. If they happen before motility is completely abolished, they may alternate with the cramps.

Patho-Anatomy.—The morbid appearances of the *spinal cord* will be first described. The membranes are usually healthy, although sometimes there are a few brown or amber-yellow stains on the pia mater immediately over the affected segments of the cord; the stellate cells in these spots are filled with pigment matter.[18] Often, through the pia mater, *grayish patches* may be distinguished here and there on the surface of the cord. On stripping off the membranes, which is easily done, these are found to have irregularly oval but, well-de-

[18] In one of Vulpian's cases (No. 2) there were several fibroid patches adherent to either layer of the arachnoid. There was hyperæmia of the pia mater of the inferior segment of the cord in another (No. 3), with some cartilaginous patches on the visceral layer of the arachnoid. When meningitis is found after death, it is almost always recent, about the cauda equina, and ascending, and secondary, being caused by the sloughs on the sacrum.

fined outlines, of an ashen or ground-glass color, and to be distributed over several columns of the same side, or confined, or nearly so, to symmetrical columns. They frequently intersect the fissures and lines of emergence of the nerves. Their dimensions are variable, generally from three to four centimetres long, by two or three broad; they may be mere grains or linear streaks. They are firmer to the touch than the surrounding tissue, and somewhat depressed and shrunken, though in a few instances they were slightly prominent (Cruveilhier, Charcot), probably indicating an early stage of development, for they were turgescent and less dense. By exposure to the air the patches acquire a rosy or salmon tint.[10] On slicing them, they prove to be conical masses, more or less deeply embedded, or rather wedged, into the white substance. They shade off imperceptibly into the healthy tissue, there being no precise delimitation. (Fig. 1.) Most commonly discrete, they are sometimes confluent, and Cruveilhier has represented this arrangement very accurately in one of his plates.

FIG. 1.

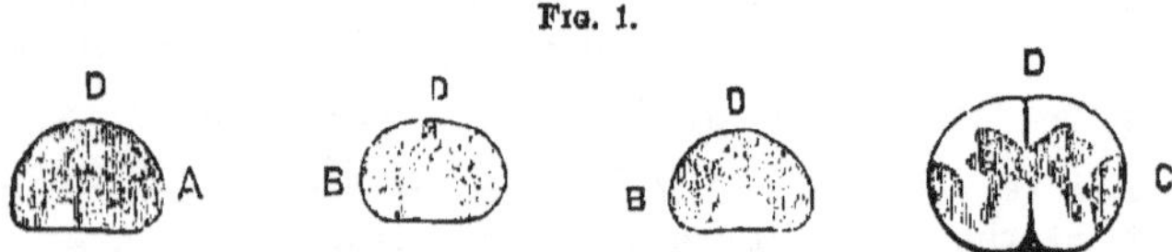

Transverse sections of the cord. D, Anterior fissure. A, Above cervical enlargement. B, B′, In the middle portion of cord. C, Three centimetres above terminal end of cord. —CHARCOT.

In the medulla oblongata the patches are met with separately or conjointly on the olivary and restiform bodies, and the pyramids, but most often on the olivary bodies. The pons varolii is the frequent seat of sclerosis, especially its inferior surface, where it appears as gray patches, with a wavy contour, disposed transversely across the median line, and may extend to the cerebral peduncles in one direction, and the

[10] If portions of the cord are placed in a solution of chromic acid, the diseased tissue becomes at first yellow, then white and opaque, contrasting strikingly with the greenish-gray tint of the healthy substance; or, as suggested by Bouchard, a camel's-hair pencil, moistened with an ammoniacal solution of carmine, may be passed several times over the surface, which is then exposed to a stream of water. The carmine will have stained only the altered tissue.

medulla oblongata in the other. In one case, the aqueduct of Sylvius was surrounded by a large patch, with processes on the pons, and towards the fourth ventricle. (Fig. 2.)

FIG. 2.

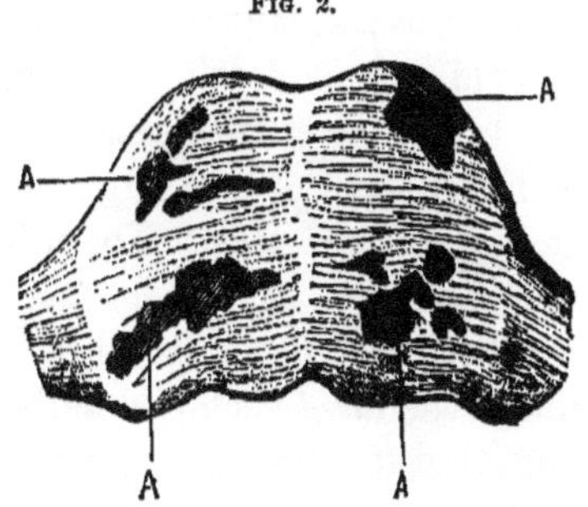

Section of the pons varolii. A, A, A, A, Patches of sclerosis on cut surface of upper portion.—CHARCOT.

The cortex of the cerebellum is scarcely ever affected, the patches of sclerosis, when present in the organ, being in the white substance, or in the corpus dentatum.

Rare in the convolutions, the patches are, sometimes, found where the gray and white substances join, but are more common on the walls of the lateral ventricles, and may extend to the intraventricular nuclei of the corpora striata. When existing in the centrum ovale, corpus callosum, and septum lucidum, they are very apparent, and are generalized. They have occurred in the optic thalami.

Though as a rule the cerebral nerves escape, in one of Cruveilhier's cases, the hypoglossal, glosso-pharyngeal, and pneumogastric were implicated; in two cases reported by Drs. Vulpian and Liouville, the optic and olfactory nerves had patches visible to the naked eye; and in one of Dr. Ordenstein's observations the hyperglossal and left motor oculi externus were distinctly sclerosed. In their configuration they resembled the patches in the pons and medulla oblongata, and in nowise could be confounded with the tissue-changes of these nerves noticed in locomotor ataxy.

With one exception, the spinal nerves have been free from sclerosis, the nerve-roots often issuing perfectly sound from the midst of a patch. At a meeting of the Société de

Biologie, July 30, 1869, Dr. Liouville stated that he had seen one case where the spinal nerves had sclerotic patches (Mémoires, etc. l. c.)

Histological Alterations and Microscopical Appearances in Different Stages.—In studying in detail the essential anatomical characters of this lesion, it is necessary to understand the modifications the histological elements undergo in the successive phases of the morbid processus. The minute anatomy of the nervous system has been accurately studied but quite recently, and though there are yet many points under discussion, certain fundamental facts have been generally accepted; these should be borne in mind to intelligently appreciate the exact nature of the tissue-changes in disease. The proper elementary structure of the nervous centres, consisting of the nerve-fibres or tubes, and ganglion-cells or corpuscles, lies embedded in, or, more correctly, is bound together by, a kind of connective tissue, which Kölliker called reticulum, and Virchow neuroglia. It is this connective framework, or gangue, which is specially interested in the lesion under consideration.[20]

If a thin section of the cord, hardened in a solution of chromic acid, and colored by carmine, and afterwards made transparent, be examined with a low power, the white matter seems at first to be entirely composed of small, regularly-rounded bodies, or disks, placed side by side, and of about the same diameter; these are the cut-nerve tubes; the small red points, looking like minute globules, in the midst of these shining translucent disks, are the colored axis-cylinders. A closer examination shows that these disks are not contiguous, but are really separated by an apparently homogeneous substance, not so deeply tinted as the axis-cylinders, and which seems to fill up, like a cement, the interspaces between the nerve-tubes. This is the neuroglia or reticulum. In the gray substance this interstitial matter is still more abundant than

[20] The connective tissue of the cord was first investigated by Keuffel (Reil's Archiv, x., 1811). Cruveilhier, in the article Apoplexie, Dictionnaire de Médecine et de Chirurgie Pratiques, ed. 1820, writes: "The extremely delicate serous cellular tissue which unites and separates the cerebral fibres, and which forms an excessively fine web. . . . ."

in the white; certain portions are wholly formed by it, as the periphery of the centrum ovale, and the central thread of the ependyma (Virchow). It is in excess, too, in the gelatinous substance around the expanded extremity of the posterior cornua of the cord, and in the posterior commissure, which, when treated with a solution of carmine, shows a nearly uniform reddish hue, while the anterior commissure, from the number of nerve-tubes which cross it transversely, is less perfectly stained. The meshes of the neuroglia are finer in the gray substance than in the white, and its texture appears more cellular or spongy. This connective framework in both substances gives support to the ramifying blood-vessels. The zone of this tissue around the peripheral portions of the nervous centres, described by Bidder and Frommann, is called the cortical layer of the reticulum; it sends processes towards the central parts of the cord, which form triangular compartments or trabeculæ, their bases being on the circumference, and apices in the gray matter. These processes divide and subdivide, forming a reticulated arrangement, whose interspaces vary in size; in the largest of these, eight or ten nerve-tubes may be lodged, while the smaller ones may contain only a single tube.

What is the histological constitution of the neuroglia? This is an open question, and micrographers are still disputing about its intimate nature. On one point they are, however, pretty generally agreed, namely, that it is not ordinary connective tissue.[21] The opinion most largely held (From-

[21] The anatomical fact of the reticulum or neuroglia, is disputed by several histologists, especially by Robin and Henle. The former asserts that it has no natural existence, but is a product of art. He holds that in a fresh state, before the cord has been subjected to the action of reagents, the spaces between the nerve-tubes are not filled by a reticulated connective tissue, but by a grayish, soft, amorphous, and finely granular matter, in the midst of which the myelocytes (the nuclei of the neuroglia of Virchow) are suspended. Alcohol and certain acids, especially chromic, cause this matter to become hard without shrinking, and it is to this property that the reticulated arrangement is due.* The answer to this is: while it is admitted that in the fresh cord amorphous matter in very

* Programme du Cours d'Histologie, 1870. Dictionnaire Encyclopédique des Sciences Médicales, 2me serie, t. i.

mann, Max Schultze, Kölliker), and which Dr. Charcot's personal observations tend to confirm, is, that it is constituted as the stroma of the lymphatic glands, after the type of the simple reticulated connective tissue of Kölliker. If, according to those authorities, thin sections of the cord, prepared in chromic acid, and colored with carmine, are examined under the microscope, the following conditions are noticed: 1. In the white substance, at those points of the reticulum where several trabeculæ meet, enlargements or knots are seen, in whose centre are nuclei, with a well-defined contour, deeply colored, and which are the myelocites of Robin, and the nuclei of the neuroglia of Virchow.[22] In transverse sections, the trabecular structure, with its homogeneous, brilliant, fibroid-looking walls, constantly anastomosing, seems to form interspaces con-

small quantity is interposed between the nerve-elements, and that in this state the reticulum is less clearly made out than in sections which have been hardened in chromic acid, it is not less true that in fresh pieces of the cord the white substance, placed in iodized serum, and dilacerated under the microscope, show unquestionably on their edges the presence of connective tissue (Kölliker, Frommann, M. Schultze); and that this tissue is more susceptible of demonstration in certain morbid conditions, in which the healthy structure becomes exaggerated, without undergoing complete transformation (Virchow). This is particularly the case in simple sclerosis, when the alteration has not gone beyond the first stage of its evolution; the effect of these reagents is limited to rendering more distinct the reticulated structure of the connective framework of the spinal cord.† Henle and Meckel have recently attempted to show, by observations which are entitled to consideration, that the chemical reactions of the neuroglia are directly the opposite of those of connective tissue. (*Ueber die sogenannte Bindessubstanz der Centralorgane des Nervensystems*, Henle and Pfeufer's Zeitschrift. Bd. xxxiv., 1869). Of course, the pathogenetic processus of sclerosis of the nervous centres is differently interpreted according to the histological theory held: one set regarding it as an exaggerated development of preformed tissue—a simple hyperplasia, as sclerosis of the lung, liver, spleen, etc.; and the other, as a lesion *sui generis*, consisting of the heterologous, or abnormal formation, of an element or tissue, differing from the physiological type of the organ; in fact, a ***heterotopical neoplasm.***

[22] Virchow, R. Die krankhaften Geschwülste, t. ii., p. 127; 1864–'65.

† Hayem. Différentes Formes de l'Encéphalite, 1868. Hayem and Magnan. Journal de la Physiologie, No. 1, 1869.
Deiter. Untersuchungen über Gehirn und Rückenmarks, p. ii., Braunsweig, 1865.

taining the nerve-tubes; while in longitudinal sections they appear to subdivide and ramify indefinitely, producing the finest meshes, which are interposed between the nerve-tubes. The interspaces between the sheaths and the tubules are filled by a small quantity of finely-granular amorphous matter.

We will now briefly examine the histological changes which are met with in those portions of the spinal cord affected with disseminated sclerosis.

According to Frommann, who had an opportunity to observe the tissue-changes at an early stage of the morbid processus, and Rindfleisch, the walls of the arterioles and capillaries are thickened by a proliferation of small, rounded cells; the cells of the neuroglia are increased in size, and their processes dilated; the nuclei are large and surrounded by a layer of protoplasm, while their number is increased, one cell containing from two to ten nuclei, either in one mass, or each with its own membrane, in which case each of these nuclei appears as a cell detached from the envelope and contents of the primitive connective corpuscle. In consequence of the progress of the nuclear multiplication, nuclei form in the processes of the connective corpuscles, and these processes grow larger as their anastomoses become more apparent. A little later the anastomotic net-work of the connective processes is filled with nuclei, and it is by the abnormal accumulation of these elements that the sites of the original cells are discerned.

The sum of these changes may be shortly stated: as an overgrowth of connective nuclei and cells, and the subsequent development of fibrillary elements of great tenuity, which may be seen on the cut surfaces, and behave like elastic tissue. These elements originate probably in the amorphous fundamental matter of the neuroglia. According to Rindfleisch, the morbid tissue is moistened by a viscous, intercellular liquid, slightly coagulable in water, and containing the nuclei and small uninuclear cells; by pressure, this liquid is forced out on the surface of a recent section, giving rise to the ependymoid formation of Rokitansky.

In consequence of the modifications of the reticulum, the nerve-elements are compressed and atrophied; the tubes lose their myelin, which undergoes granular disintegration; the axis-cylinders become paler and very translucent, with, frequently, a longitudinal striation, or have a finely-granular aspect, which Frommann regards as the beginning of the molecular destruction. Around the atrophied nerve-tubes a large number of globules are seen, formed by a coagulated fatty matter. In proportion as the atrophy of the nerve-elements advances, the fibrillary tissue is developed, and gradually assumes the characters of genuine connective tissue. Regressive transformations now take place, fatty granulations are deposited in the interstices of the tubes, and amyloid corpuscles are produced at the expense of the small uninuclear cells of the intercellular viscous liquid. The absorption of this liquid, favored by the contraction of the connective elements, marks the last stage of the processus. The diseased tissue, deprived of its juice, is composed solely of connective fibres immediately contiguous, whose unequal shrinkage gives rise to partial distortions. Finally the whole mass becomes of a grayish fawn-color, often closely resembling the tint of ground glass.

Charcot has described these histological changes in the several stages of the processus more fully. On placing under the microscope a properly-prepared thin transverse section of the diseased cord, several concentric zones are recognized; these correspond to the different phases of the lesion. *a.* In *the peripheral zone*, the trabeculæ of the reticulum are thickened; the nuclei of the enlargements of the reticulum are swollen; the nerve-tubes are atrophied at the expense of the medullary sheath, the axis-cylinder being whole. *b.* In *the second, or transitory zone*, the size of the nerve-tubes is still less—indeed, in many the myelin has disappeared, the axis-cylinder only remaining, which may be hypertrophied. The trabeculæ are more transparent, their contour less definite, and at certain points they are replaced by bundles of long, delicate fibrillæ, analogous to those which are characteristic of common connective tissue. These are developed at the expense of the meshes which contain the nerve-tubes, and the consequence is that the reticulated aspect of the healthy neuroglia has a tendency to disappear. *c.* In *the central zone*, that is, in the midst of the sclerosed patch, all traces of fibroid reticulum have gone; there are neither cells nor trabeculæ; bundles of fibrils fill the alveolar interspaces, from which the myelin has entirely disappeared; the axis-cylinders are atrophied to such a degree that it is hard to distinguish them from the newly-formed fibrils. The persistence of these cylinders in the midst of the tissue which has undergone fibroid substitution, is, Dr. Charcot thinks, peculiar to disseminated sclerosis.

The characters of the fibrillary tissue are particularly well exhibited in longitudinal sections. It is seen to be composed of parallel fasciculi, formed of delicate, opaque, smooth fibrils, rarely subdividing and anastomosing, but frequently interlacing, after the manner of felting, and which are colored by carmine. These characters enable us to distinguish it from the axis-cylinders, which are usually larger, transparent, and do not ramify. (Fig. 3.)

Fig. 3.

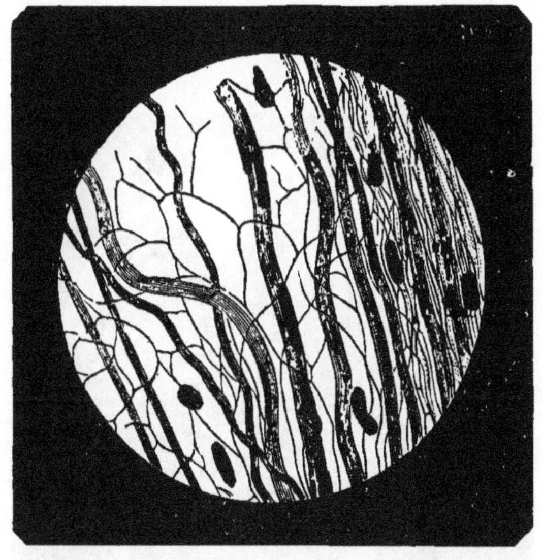

Recent preparation from centre of a patch of sclerosis of the cord, colored by carmine and dilacerated. In the centre a capillary vessel with numerous nuclei. On either side axis-cylinders, some large, others quite small, deprived of their myelin. The capillary and the axis-cylinders, strongly colored by carmine. The latter are perfectly smooth, and have no ramifications. Between the axis-cylinders, delicate fibrils of recent formation are seen; on the right nearly parallel to each other, while in the centre and on the left, they form a sort of net-work, resulting from their interlacement, or anastomosis. They may be distinguished from the axis-cylinders by their diameter, which is much less, by their numerous ramifications, and by not being colored by carmine. Scattered nuclei are also visible; several are in connection with the fibrils, and others are of irregular form, from the action of the ammoniacal solution of carmine.—Charcot.

They may be distinguished, also, from the fibres of the reticulum by these being thicker, shorter, and always having branchlike processes. Finally, they differ from the elastic fibres by their swelling up from the action of acetic acid, and forming a transparent hyaline mass, which is not the case with the elastic fibres. In the midst of this fibrillary tissue there is always a considerable number of corpora amylacea.

The vessels which traverse the patches of sclerosis, undergo alteration. In the peripheral zone, the walls of these vessels are thickened, and contain more nuclei than in the normal state; nearer the centre the nuclei are more numerous, and the lymphatic sheath is replaced by several layers of fibrils like those developed in the thickness of the reticulum. Owing to the concentric hypertrophy of their walls, the calibre of the blood-vessels is lessened.

There are certain changes which can be best or only studied in the recent state. In the centre of the sclerosed mass, globules or granules of a fatty kind are nearly constantly met with. These elements have two chief aspects: in the one the masses have sinuous, dark outlines, with, like the myelin, a double contour; the other are true fatty granulations, sometimes isolated, at others agglomerated, so as to form granular bodies. (Fig. 4.)

FIG. 4.

Patch of sclerosis from cord in a recent state. *a.* Lymphatic sheath of a vessel distended by large fatty globules. *b.* Vessel cut across; the lymphatic sheath is separated from the tunica adventitia by an empty space, the fatty globules which distended the sheath having disappeared. *c. c.* Little collections of fatty globules scattered over the preparation.—CHARCOT.

These drops of myelin and fat granules may infiltrate through the meshes of the reticulum, and spread themselves abroad; they never occupy the centre of the patch, for there the fibrillary metamorphosis and the destructive process as regards the nerve-tubes are ended; but they are seen about its periphery at those points where the medullary sheath has almost disappeared, by the compression first of the trabeculæ, and afterwards of the fibrillary fasciculi, which gradually invade and occupy the alveolæ. The nerve-tube is finally represented by the axis-cylinder alone. The accumulation of the medullary or fatty globules and the destruction of the medullary sheath are contemporaneous, and when the one is accomplished the other ceases to occur. We may then infer that the medullary and fatty corpuscles are detritus resulting from the disintegration of the nerve-tubes. Dr. Charcot believes that the absence of these fatty granules in the centre of the patches is due to their absorption, and this view seems sustained by the fact that within the lymphatic sheaths of the arterioles those fat-corpuscles are sometimes so abundant as to notably increase their volume, and are visible by the naked eye in the form of white lines on the gray ground of the patch. The changes in the vessels before described are not a primitive lesion, or an essential part of the degenerative processus, but simply a consecutive fatty infiltration of the lymphatic sheaths.

These morbid changes are also met with in the gray matter; the ganglion-cells, however, are not the seat of the nuclear proliferation observed in the connective cells, but undergo a peculiar alteration known by the name of *yellow degeneration*, from its ochrey tint: they can no longer be colored by carmine, as in the healthy state; and finally, all the constituents of the cell become atrophied.

The consecutive phases of the morbid processus may be summarized as consisting: 1. Of the initial fundamental fact —*proliferation of the nuclei, with concomitant hyperplasia of the reticulated fibres of the neuroglia;* and, 2. Of the *secondary, consecutive atrophic degeneration of the proper nerve-elements.*

Differential Diagnosis.—This affection has sufficient individuality to secure, in most instances at least, its ready recognition, notwithstanding resemblances between its several forms and other diseases of the nervous system. The disorder with which it has been heretofore invariably confounded is, as has been stated, paralysis agitans, or shaking palsy. Besides a pathogenetic distinction founded upon the sure basis of morbid anatomy, there are differences of clinical history which should forbid confusion or mistake. Paralysis agitans is chiefly met with in persons of declining years (Trousseau). Sauvages says, "While chorea, or *scelotyrbe sancti viti* attacks the young, ballismus, or *scelotyrbe festinans*, attacks those of advanced life;" and Saunders observes, "Age is of primary importance, both in causing the affection and aggravating it." Disseminated sclerosis of the cerebro-spinal centres is a disease of adult life. There is some parallelism in the invasive stage of the two disorders, the approach of both being often so imperceptible that it is difficult for the patient to fix the exact period of commencement. In both there are crawling sensations and numbness, and sense of diminution of muscular power; but in the neurosis these are almost constantly felt in the arms, while in the organic disease they are almost as constantly experienced in the legs. In paralysis agitans, a tendency to trembling is an initial symptom, and precedent to the paresis; in sclerosis, on the contrary, tremor invariably follows paralysis, limb by limb. The muscular weakness of paralysis agitans begins in one arm, or both, and then extends to the lower extremities; and when it does pass into true paralysis, which is very rare, it is in the final stage; in diffuse cerebro-spinal sclerosis, one or both lower limbs are first attacked with paresis, and, sooner or later, become completely deprived of all motor power, rendering the patient perfectly paraplegic and unable to walk. Hence the gait in the two diseases is eminently diacritic. A patient with shaking palsy in its developed stage, after the customary balancings and oscillations of the body, starts with the head and trunk bent forward, on the toes or forepart of the feet, with short, quick steps, and, to maintain the centre of gravity, thus displaced, goes trotting and hopping along, at almost running speed, with one or both arms and wrists semiflexed and closely pressed to the sides; he seems, as Trousseau observes, to be running after himself. This festinating, procursive gait is pathognomonic. Nothing of the kind is seen in cerebro-spinal sclerosis; the gait is that of paraplegia, according with the degree of palsy, and will be presently more fully described. *Tremor* is a prominent and common symptom of the two affections, and from this has arisen much of the confusion which led writers so persistingly to associate pathological species genetically distinct; for when this sign is analyzed and comparatively studied in the two disorders, great differences, not only of development but of degree, become evident. One of the earliest phenomena of shaking palsy, in disseminated sclerosis it is invariably consecutive to the motorial troubles. In the former disorder, trembling is incessant, and is but little modified whether the patient be at rest or in motion, and in the developed stage is scarcely interrupted by sleep; while in the latter, it is never spontaneous, but is always provoked by, or follows upon, any muscular movement; and though it may be started by emotion, it will be found, on observation, that this deviation from the rule is only apparent, for some change of position of a limb, or of the head or trunk, is always involved. Nystagmus, so constantly present in sclerosis, is never met with in paralysis agitans. The peculiar slow, scanning articulation of a sclerotic is very different from the embarrassed, indistinct utterance noticed in paralysis agitans. The intellect in paralysis agitans is generally unaffected until near the close, while in diffuse cerebro-spinal sclerosis it is mostly weakened from an early period. The characteristic deformities, described first by Parkinson, and recently by Charcot,[23] which occur in shaking palsy, cannot be confounded with the permanent contractions of late muscular rigidity, so constant in the terminal period of diffuse sclerosis.

There are some symptoms in common between the ordinary forms of multilocular sclerosis of the anterior, or antero-lateral, columns of the cord and locomotor ataxy (posterior sclerosis) in the forming stages of the two disorders. In both there are often tinglings, and occasional numbness, and ready fatigue after slight muscular exertion. In locomotor ataxy these are, however, very often accompanied by ocular troubles, as weakness of sight, defective accommodation, strabismus, ptosis, or double vision. In the spinal form of disseminated sclerosis these are wanting, and when they happen in the cerebral, or cerebro-spinal form, they are always persistent, while in locomotor ataxy they are generally temporary. The terrible boring, gnawing, or lancinating, pains of the ataxic, so frequently preceding marked motorial troubles, are rare in disseminated sclerosis. The course and physiognomy of the two diseases in their developed state, when one does not com-

[23] In the hand, the contortion is very characteristic. At first, the phalanges are partly flexed on the metacarpal bones, and drawn toward the thumb, as if the patient were holding a pen, or about taking a pinch of snuff. After a while this condition is aggravated, the first phalanges of the fingers are closely and tightly bent, while the second are in a state of permanent hyperextension; on the first and the third in a slight degree, of flexion on the second (Ordenstein).

plicate the other, are too distinct to allow of any confusion. The characteristic hitchy, catching-up gait of the ataxic, resulting from impaired muscular adjustment, has nothing in common with the paraplegic shuffling of Charcot's disease. Even in the paretic stage of multilocular sclerosis, the gait is distinctive; the lifted leg is brought forward, not in a direct antero-posterior plane, as in the physiological act of walking, but the foot describes an arc of a circle, and from within outward, at first; and then, having reached the extreme point of the curve, it is thrown inward, coming to the ground with a flap, the sole striking at all points at the same time. With this excentric curvilinear projection of the foot, there is a cadenced oscillation, an exaggerated alternate semi-rotation of both halves of the pelvis (Jaccoud). Early paresis passing, certainly, and more or less speedily, into paralysis, characterizes multilocular sclerosis; while paraplegia is always a very late phenomenon in ataxy, and in nowise a necessary one, so long as the tissue-changes are limited to their own district. When the posterior columns of the cord are invaded by the sclerotic patches, already existing in the anterior, or antero-lateral, then the signs of the two affections will coexist, but the phenomena proper to each can generally be separated by careful examination.

Course and Duration.—The course of this affection is progressively invasive, whether it begins in the lower or upper extremities, or in the head. Muscular enfeeblement slowly but certainly increases, until all the voluntary muscles are more or less disabled. Tremor, at the outset slight, partial, and occasional, comes to be violent, spreads over the whole body, and is started on the least voluntary movement, or by any emotion. Convulsive agitations of the extremities, spells of rigidity, and permanent contraction of the muscles succeed; nutrition troubles set in, and prepare the way for some acute intercurrent disease which is commonly fatal. But there are a few exceptional instances in which the advance was not continuous. In one, there were repeated pauses and recommencements before the disease became regularly progressive (Vulpian). Leo reports a case in which the palsy, at first uni- and then bilateral, disappeared for a while altogether. One patient improved so much under the use of strychnine, that he could stand without support, which he had not been able to do for some time; and this continued for two months (Zenker). Another recovered of the paraplegia, and rectal and vesical paralysis, and during six months could walk alone, when the symptoms started afresh (Bourneville). Dr. Charcot relates the case of a woman, who for two years was obliged to remain almost completely motionless, the limbs being permanently contracted, with scarcely a few short intervals of relief; at the end of that time she became able to walk about and attend to her household duties; when, after a fit of convulsive hysteria, all her limbs and the muscles of the trunk became rigid, and remained persistently so to her death, nine years later (*Gaz. Hebdom.*, February, 1865). In two cases there was decided amendment on the appearance of the catamenia, and which ceased on their stoppage. Another remarkable case of repeated pauses, improvement, and aggravation, is reported by Vulpian. During an attack of small-pox, both paraplegia and contraction ceased in the eruptive stage, and after convalescence all signs of the disorder had vanished. This lasted for three years, when, after a fright, which suppressed the menstrual flow, slight paresis was felt; this passed off, however, on the reëstablishment of the catamenia. Three years subsequently this patient had an attack of jaundice; the motor troubles recommenced, and, after some fluctuations, pursued a regularly onward course.

The average duration may be stated at from eight to ten years. In seventeen fatal cases, it lasted for two years in 2, five to ten years in 9, thirteen years in 2, seventeen years in 1, twenty years in 1, and twenty-four years in 1. The age of death in these cases was: from twenty-five to thirty, 2; from thirty to thirty-five, 5; from thirty to forty, 4; from forty-one to forty-five, 2; from forty-six to fifty, 2; from fifty-one to fifty-five, 1.

Causes and Pathogeny.—It is essentially a disease of adult life. In the eighteen cases collected by Bourneville, fourteen were between twenty-six and thirty-six years of age; and of these, fifteen were females; but it must be borne in mind that, of the physicians who have busied themselves with the study of this affection, nearly all were attached to institutions special to women, the Salpêtrière, for example. In thirteen cases in which the occupations were known, there were: 1 physician, 1 theological student, 1 male teacher, 1 female teacher, 1 pianist, 3 seamstresses, 3 domestic servants, 1 truck gardener (female), 1 flower-woman. Several would seem to have suffered from previous nervous disorders, as hysteria, hemicrania, etc. Excess in alcoholic drinks is mentioned in one case. A fall immediately preceded the onset in another (Vulpian). In three instances it appeared during pregnancy. Once it was developed after violent fatigue. Exposure to damp cold may probably be named as one of the few known causes; it is particularly mentioned by Parkinson as giving rise to paralysis agitans. There is no evidence to show the influence of heredity; the father of one patient is reported as having had tremor of the head at fifty.

It must be allowed that the *pathogeny* of this affection, as well as that of other kindred diseases of the great nervous centres, is obscure and unsettled. May we not possibly find a truer interpretation of their nature by admitting their connection with and dependence on some latent but distinct condition of the constitution of the body, which favors the development of this peculiar form of morbid action? That they are simply varieties of chronic inflammation of the brain and spinal cord cannot well be admitted. They are more than this. The common assigned causes seem insufficient for their genesis without some impalpable preëxisting tendency. May they not be like Bright's and Addison's diseases, local expressions of a general state, in which a specific diathesis enters as an essential factor? When sclerosis of the posterior columns of the cord (locomotor ataxy) comes to be treated of, this theory of the pathogenesis of the class of disorders under consideration will be more amply examined, and the evidence in its favor more largely set forth. Should this view gain favor at any time, its practical application is clear, and it will necessarily be followed by an entire change in the therapeutics of these disorders.

Prognosis.—No instance of recovery has been reported. In the actual state of our knowledge, it may be said to be incurable, and progressively fatal.

Treatment.—Electricity, blisters, counter-irritants, ergot, arsenic, and belladonna, have all been used, without good effect. Chloride of gold and phosphate of zinc aggravated the symptoms (Charcot, Vulpian). Strychnine temporarily modified the tremors in five cases. Nitrate of silver appears to have done the same, but it does harm if there is muscular contraction, or clonic spasms. One case is reported as improved by tonics and sulphur-baths. Dr. Pennock was benefited by hydrotherapy. The constant galvanic current should in all instances be fairly tried. If the writer's suggestion as to the diathetic character of this and similar disorders should prove correct, it may be possible, at least in the early stage, to arrest the progress of this affection, by a method of treatment founded on this theory. This practical point will be more particularly discussed and applied in a subsequent article.

APPENDIX.

Case I.—*Spinal Form.* (Reported by J. C. Morris, M. D. Microscopical Examination by S. Wier Mitchell, M. D. Transactions of the College of Physicians of Philadelphia, 1868.)

Dr. Caspar W. Pennock, of Philadelphia, of large frame and muscular development, began, in the winter of 1842–'43, when about forty years of age, to feel numbness and a sense of heaviness in his left leg. The sensation was that of a tight band around the leg, three or four inches in width He had been laboriously occupied during the summer with professional duties, which severely taxed both his mental and physical powers. During six years, in spite of counter-irritation to the spine, galvanism, and hydrotherapy, (which latter seemed to have some temporary effect), the symptoms increased, extending to the whole limb, and then appeared in the right lower extremity; and in 1849 he had become gradually paraplegic. In 1853 he was attacked with apparently phlegmasia dolens in the lower limbs. It ran its course in three weeks, leaving him ever afterward unable to walk, and he was entirely confined to the sitting or horizontal position. In time the paralysis reached the upper extremities, beginning in the left arm, and then spreading to the right, and becoming

3

so great as to prevent him during the last ten years of his life from feeding himself. Sensibility of the paralyzed limbs was somewhat lessened, but never entirely lost. Five years before death there was temporary paralysis of the bladder. His intellect was uninterruptedly clear and bright to the latest period of life, and he took a deep interest in every thing. Death from acute intercurrent disease in 1867, twenty-four years after invasion. Course of disorder regularly progressive. There was total loss of voluntary motor power below the neck. Reflex action intact.

Autopsy.—The cerebrum, cerebellum, and pons were without notable change. At the upper part of the posterior face of the medulla oblongata, in the mouth of the fourth ventricle, lying between the restiform bodies and posterior pyramids, there was a small, irregularly-rounded concretion, of two lines in diameter.

The cervical and dorsal regions of the spinal cord presented in the fresh state, to the naked eye, a series of gray translucent spots, of irregular form, and sometimes almost transparent. They were all a little sunken below the level of the surface of the cord.

The microscopic examination of these spots, which lay chiefly in the white substance, showed—1. Total absence of nerve-tubes and nerve-cells; 2. Finely-granular matter, molecules, and small globules of fat in great quantity; 3. No granulation-corpuscles; 4. Numerous fibres. The vessels of the cord were everywhere altered by fatty deposits, which were particularly conspicuous in the neighborhood of the spots. No vessel was detected in the altered tissue. Excepting in a small spot which invaded the right posterior horn of gray matter at the level of the first cervical nerve, and a like spot on the left side in the same situation, the change respected the posterior horns and columns of the cord. The lateral columns were extensively affected, the spots, as a rule, lying between the anterior nerve-roots, and the central line of the lateral columns, involving, therefore, most largely, the parts nearest the anterior nerve-roots (see Figs. 6 and 7). In three places the changes passed across the anterior columns. From the seventh dorsal to the eighth dorsal nerve, a large spot extended across both the anterior columns; its depth and form are seen in Fig. 5. About the level of the tenth dorsal, and at the second lumbar nerves, two spots were found crossing on to the right and left anterior columns respectively, but not involving the whole width of either. Below these points no spots were found in any part of the cord. The most extensive lesions were in the lateral columns of the cervical cord. Their surface extent is shown in Figs. 6 and 7. Their depth and the parts affected in the interior of the cord are seen in Figs. 8, 9, 10, 11 and 12.

Fig. 5.

P

A

Between seventh and eighth dorsal.

In the dorsal region the spots were abundant, but less numerous than above, about half as many for equal areas of tissue. In many portions both of the cervical and dorsal regions, the changes involved the point of entry of the anterior nerve-roots and the gray matter of the anterior horns (*see* Figs. 10, 11, and 12). The central canal was distinct throughout its course.

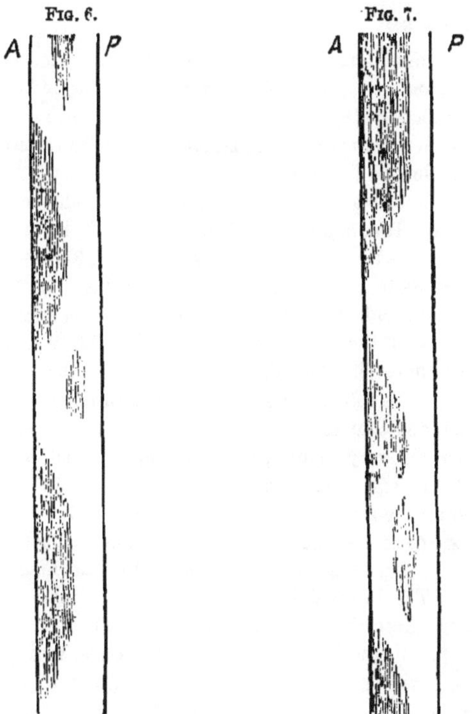

Right and left lateral surface view of the lesions of the cervical cord.

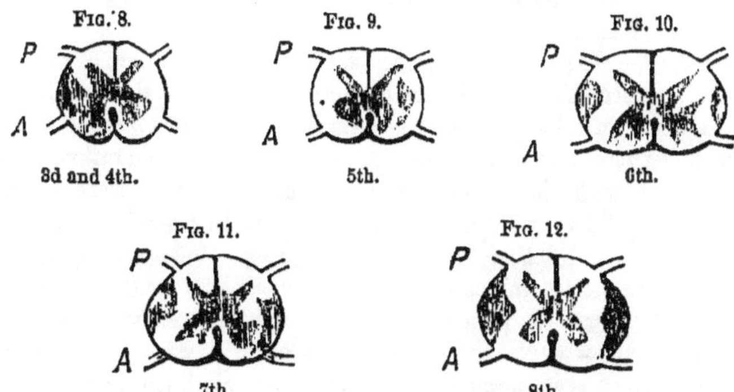

Cross-sections of cervical cord, showing lesions at the level of the nerve-roots, indicated by the numbers.

Case II.—*Spinal Form. No sclerotic patches visible to the naked eye on the examination of the cord. The microscope shows the characteristic lesion.*—(Charcot; reported by Bourneville.)

C. C., female, aged forty-one, seamstress, widow. Intellect very feeble, and memory almost gone; not able to give any history of her case. Entered hospital February 18, 1868. Face pale, pupils natural; vision a little weakened; speech embarrassed, and accompanied with tremor of the lips; stiffness of muscles of the neck; hands cold and purple. In a state of repose, nothing abnormal in upper extremities. Rigidity of the shoulder and elbow-joints. Holding her snuff-box in the left hand when she attempts to open it with the right, it is at once affected with tremor. When she attempts to carry a glass of water to her mouth, holding the vessel in the right hand, she always fails after repeated efforts. Muscular power diminished in the right upper limb; sensibility to pinching and pricking appears natural. The thighs are flexed on the pelvis at a right angle; the legs on the thighs; the lower extremities regarded as a whole are in adduction, and that to such a degree that the knees can be separated from each other but very slightly, without causing great pain. It is possible to extend the legs somewhat without causing pain to the patient. If, after extending the right leg as much as possible, it is raised from the bed, the foot begins to tremble; and, if at this time the sole of the foot is tickled, there is produced a sudden spasm in the whole limb, as from an electric shock. The different kinds of sensibility natural in both lower limbs. Although from the mental state of the patient the commemoratives were wanting, the diagnosis, made with some reserve, was spinal disseminated sclerosis, based on her condition when admitted into the hospital; the paraplegia and permanent contraction of the lower extremities, the tremor on voluntary movements, the spinal epilepsy, etc. The absence of trembling of the head and of nystagmus seemed to show that the brain was not implicated.

Autopsy.—No patches of sclerosis were visible in any part of the cord; but on microscopic examination of sections of different portions of the cord, the existence of sclerosis was made certain by the histological changes present. There was considerable connective-tissue proliferation, in certain districts, with fibroid substitution, but the nerve-tubes were yet generally intact. "Are we," asks the reporter, Dr. Bourneville, "from this case to conclude, by those lesions not being appreciable to the naked eye, that there may be several varieties of sclerotic patches? or that the change was yet in its earliest period?" This latter hypothesis seems, from the histological appearances, to be the correct one. Besides, the same thing has been already seen in sclerosis of the posterior cords (locomotor ataxy), as in the instances reported by Dr. Gull, of London, and Drs. Charcot and Bouchard, in which no tissue alteration apparently existed, and the microscope revealed the characteristic changes.

Case III.—*Cerebro-Spinal Form.*—(Charcot; reported by Bourneville.)

H. B., female, aged forty-one, admitted January 6, 1868. Health delicate in youth; after fourteen, when menstruation began, good. Married at twenty-three; has had three children. At the age of twenty-eight, during second pregnancy, had a fall. When about thirty, after a long walk, first felt weakness in the legs; this gradually increased, so that in two years she could not walk without aid; at this time there was incontinence of urine and fæces. Under a tonic treatment and sulphur-baths, the paralysis of the sphincters disappeared, and her lower limbs became so much stronger that she could walk without support, but could be easily thrown down. This improvement lasted for six months, when the paraplegia returned and regularly progressed. The lower limbs felt weighted and stiff. During her third pregnancy walking became impossible; she had to be lifted in and out of bed. About nine years after onset she began to have a sensation of fatigue in both upper extremities, and in the same degree. The rigidity of the lower limbs increased. Her condition on January 7, 1868, day after her admission to hospital: General health and nutrition good; constipation; menstruation unaffected. Rigidity of lower extremities, which are in a state of adduction; rigidity most marked on left side; motility completely abolished; sensibility to pricking, pinching, tickling, and cold and heat, natural; the latter, perhaps, slightly increased; no pains in limbs, no formication or numbness, but a constant feeling of fatigue and heaviness; no palsy of sphincters. Severe pain, almost continuous, but with exacerbations, in the lower lumbar spine and sacrum; small bed-sore over sacrum.

Both upper extremities are weak, the left more than the right. When quite still, no tremor noticeable; but so soon as she attempts to carry any thing to the mouth, there is trembling of the left side, and the head is at the same time agitated.

There had been for some time, exactly how long the patient was unable to say, weakness of sight, greater in the left than the right eye, with occasional diplopy; no periorbitar pains; no ocular delusions.

January 17.—Pains around the base of the chest, hindering respiration, and more marked on the right side. Shooting-pain spells, during which there is slight congestion of the face; the neuralgic points are: 1. To the right of fifth dorsal vertebra; 2. Below the right breast; 3. Slight hyperæsthesia at epigastrium.

January 31.—Redness in the metacarpo-phalangeal and finger-joints, with pain, increased on motion. Sacral sore larger. Rigidity of lower limbs, as well as tendency to adduction, lessened. Severe pain at times in instep, compared to the part being squeezed in a vice.

February 5.—The lower extremities, which on the day previous were without any rigidity, are drawn up, and to effect extension requires much effort, particularly on the left side, and they become immediately again semi-flexed. On pinching the skin of the thighs and legs, and tickling the soles of the feet, the legs become rigid and bent on the thighs, especially the left, and reflex movements are exerted, which throw the limb upwards. Application of cold produces the same phenomena, but to a less degree. The greater the degree of flexion, the less the tendency to adduction. When the sole of the foot is tickled for some time, tetany is induced in the limb, which lasts for some time after the excitation has ceased. Occasional subsultus in the lower limbs. After a while, the rigidity and contraction pass off, and the position of the limbs becomes natural. A constrictive pain around the hips, from the sacrum to the pubis, complained of.

In the upper extremities a feeling of uneasiness is experienced from time to time, when the hands cannot be used, and there is tremor on voluntary movement. About this time, ascending spinal meningitis from the sacral slough set in. On February 16th, the pulse-rate was 100 to 120; respiration, 36 to 40; body heat, 39°5 C. to 40°1 C. February 20th.—Involuntary stools; lower extremities flaccid and handling, and kneading them, even, do not cause contraction; tactile sensibility natural; temperature of lower extremities less than of upper; pupils normal; sight a little dim.

On the 4th of March pneumonia set in, with diarrhœa and disturbed digestion. Up to this time no headache had been felt, but now, when the head is moved, there is a deep-seated sensation in the upper part of the occipital region as if a blow had been received there, and this lasts for an hour at the time. Memory good; no disturbance of ideation; no feeling of heaviness, or of increase or diminution in the size of the head; no stiffness of the neck.

March 23.—Contactile and thermic sensibility perfect and without retardation in the lower limbs, which are relaxed; a little stiffness in the knees; some tendency to adduction; no voluntary movements, even of the toes; reflex movements exaggerated; on tickling the sole of the foot, the leg becomes flexed on the thigh, and the thigh on the pelvis; and if the excitation is continued for any time, a series of convulsive movements is provoked in the limb, which continues bent. During the night involuntary muscular startings in the lower extremities, in the direction of flexion, are frequent.

There are spells of increased weakness of the arms; the trembling of head worsens; new sloughs form; the lumbar pains are more intense; evacuations involuntary; diarrhœa; dilatation of the pupils, the right more than left; objects placed at the right of the eye appear double with both eyes, but are correctly seen by the single eye. The lung-trouble becomes developed, and the patient dies April 6, 1868.

Autopsy.—Patches of sclerosis in the cerebral hemispheres and pons varolii. In the spinal cord they are scattered throughout the whole length, and in every part except the posterior columns.

The association of fasciculated sclerosis of the posterior columns of the cord with disseminated sclerosis of the anterior and antero-lateral columns, is shown in the two following cases, reported by Friedreich in his paper on "Atrophic Degeneration of the Posterior Columns of the Spinal Cord" (locomotor ataxy), published in Virchow's Archiv. fur Pathol. Anat. u. Physiol., u. für Klin. Medicin., 1863, pp. 419–433.

CASE IV.—(FRIEDREICH.)

J. S., female: first felt weariness, then weakness of right leg at sixteen years of age; and at the same time severe shooting, intermittent pains in the lower extremities. When about twenty, she noticed increasing weakness of the right arm, and soon after of the left, while the paralysis of the inferior extremities grew worse. Nine or ten years later, severe pains were felt in the ends of the fingers, but more in the right than the left hand; about this time there were convulsive startings in the muscles of the leg, particularly while in bed, which would cause involuntary flexion of the legs on the thighs, and the thighs on the pelvis. Eleven years afterwards the head became tremulous as soon as it was raised from the pillow, and there was bilateral nystagmus, when any object was steadily looked at. Speech was embarrassed to a degree to be almost unintelligible (æt. 21); no headache, but vertigo; spasmodic contractions of lower limbs; ataxia. Death from typhoid fever, fifteen years after onset. Sensibility and vision had remained intact. At the autopsy, sclerosis of all the columns of the cord was found.

CASE V.—(FRIEDREICH.)

N. S., female: was seized with feebleness and lancinating pains in both lower extremities at the age of seventeen. Towards twenty the paraplegia increased, and a sense of heaviness in both arms was felt, but no pain. At twenty-six, speech became affected, and there was stuttering. The pain-spells in the lower limbs recurred from time to time, until her entrance into the hospital, when she was twenty-eight years of age. During the two previous years there had been occasional spasms of the peroneal muscles, and within a year severe frontal neuralgia, lasting sometimes for a whole day. On the day she came to the hospital, her speech was almost unintelligible; voluntary motion of arms difficult and uncertain, so that she could not button her clothes, or thread a needle, or squeeze an object only after many attempts; there was small control over the muscles of the arms; she could neither stand nor sit; electro-sensibility and contractility natural; in the lower extremities, strong currents, which induced violent contractions, caused less pain than in the upper; cutaneous sensibility, tested by æsthesiometer, everywhere good. Special senses natural; slight nystagmus; no paralysis of the muscles of the face; right cyphosis and scoliosis of dorsal spine. Death from typhoid fever a few months after admission. Examination of the spinal cord showed extensive sclerosis of the posterior columns, and the posterior nerve-roots were atrophied. In a segment of the cord adjacent to the lumbar enlargement, although the naked eye could detect no lesion of the lateral columns, the microscope showed connective proliferation analogous to that in the posterior column; there was fibrillary tissue, with some atrophy and decided varicosity of the nerve-tubes. In the pia mater of the cervical region and upper part of the medulla oblongata, there was intense yellowish-brown pigmentation, though no alteration of the posterior columns could be made out at these levels. [It does not appear that those portions were examined microscopically; had they been, the essential tissue-changes would, most likely, have been seen.] On exposing the fourth ventricle the ependyma of the inferior half of the rhomboid fossa, near the cord, was found to be much thickened and hard [a patch of sclerosis probably?], while the upper part was normal.

Case VI.—*Cerebro-Spinal Form, with Atrophy of Optic Disks of both Retinæ.*—(Magnan. Mémoires de la Société de Biologie, Paris, 1869.)

M. S., female, aged thirty-four; admitted July 6, 1869. At thirteen years of age (1848), had an attack of typhoid fever, which lasted six weeks. During convalescence the sight began to grow dim, and she soon became totally blind. The general health was good, and her intellect unimpaired, until the beginning of 1867, when trembling of the hands and arms was noticed whenever any movement requiring nicety or precision was attempted. The tremor gradually grew worse, and extended to the lower extremities. During the eight months previous to above date, walking was very difficult, she was confined to her chair or bed, and was unable to feed herself. Sensory troubles now appeared, pain-spells occurring in different parts of the body, particularly on the right side. While the patient remains sitting, and is perfectly quiet, there is nothing peculiar in her appearance; but, the moment she is spoken to, and she turns to answer, on moving her head, it is seized with an irregular, interrupted tremor; there is nystagmus; and the muscles of the trunk are agitated, as if by a series of shocks; the arms and hands too are tremulous, to a degree to prevent a glass of water being carried to the lips without spilling the contents. When standing, or on attempting to walk, the legs begin to tremble, and then sudden, irregular, spasmodic contractions of the muscles occur, which are quickly propagated to the muscles of the trunk, setting the whole body convulsively shaking. Speech is drawling, hesitating, and slightly thick. Darting pains and cramps are felt in the legs, sometimes in the arms, particularly of the right side; and the pains extend to the back. In the right cheek there is severe pain in the course of the branches of the facial nerve, and especially in the direction of the inferior dental and frontal branches.

At times a disagreeable sensation of burning is felt in the cheek and legs, with tinglings along the spine. On several occasions there has been a feeling of heat in the belly, and, within a few days of date, a tympanitis was developed, which seems to be passing off.

The intellect is weakened, but there is no special delusion. Sight is entirely lost. Ophthalmoscopic examination shows: in the right eye, optic disk oval, and of a pearl-white tint, with the vessels very small; in the left, optic disk white, contour well defined, and both arteries and veins of less than natural size.

REMARKS.—The point of special interest in this case, reported by Dr. Magnan, is the probable occurrence of sclerosis of the optic nerves many years previously to the development of disseminated cerebro-spinal sclerosis, which was the diagnosis. There is likely tissue-change in the facial nerve. When, Dr. Magnan remarks, similar facts are more numerous, and we have had larger opportunities of studying simultaneous alterations in both the peripheral and central districts on the one hand, and, on the other, extensions of the morbid processus from the centres, primarily affected, to the periphery, we shall be better able to understand the intimate relations between these several localizations.

The following is an abstract of Oppolzer's case, already referred to (page 5). Anatomically it is an example of multilocular sclerosis, while clinically it differs from all the other collected cases by the persistence of the trembling when the patient was at rest.

CASE VII.—*Spinal Form.*—(OPPOLZER. Canstatt's Jahresbericht, 1861, vol. iii., p. 78, from Spital-Zeit.)

Male, aged seventy-two, of diminutive stature, admitted to the clinic June 20, 1860, with violent trembling, which hindered him from using his hands. He never suffered from any serious illness until 1848 (aged sixty), when, during the bombardment of Vienna, he had a severe fright. He was carried to his home; and had scarcely recovered from the attack, when, a bomb bursting near the house, brought on another. A few hours afterward, on trying to take food, he found himself unable to use his hands, for, on attempting it, they began at once to tremble violently. After a while his lower limbs commenced to tremble, but less violently, and he could still walk. In spite of treatment, the disorder grew worse. The *trembling persisted even when he was at rest*, and involved other muscles. Paralysis (paraplegia?) came on. After a few years he was unable to stand erect, and as soon as he made the attempt, there was an irresistible tendency to fall forward, so that, to avoid toppling over, he was obliged to lay hold of any near object, or to walk hurriedly. His intellectual faculties and his senses had slowly but progressively diminished. Tea, coffee, and spirituous drinks, increased the trembling. The agitation of the lower limbs was more marked in the evening, and when the patient had walked during the day. Six months previously to admission, the sphincters became paralyzed. Five weeks before, after a severe attack of vertigo, he dropped down suddenly, and was unable to rise, but did not lose consciousness. Since that time emaciation had rapidly increased; the patient can stand and walk for a very short time only, and with very great effort.

Condition on Admission.—Great emaciation; earthy tint of skin, which is covered with numerous epithelial scales; perspiratory secretion increased on face, but seems lessened in other regions; temperature of skin lower than natural. Muscles of face, tongue, neck, and upper limbs are affected with violent trembling, which *never ceases during the waking state*, and is completely suspended only during profound sleep. The lower limbs shake periodically only, and when there is a general exacerbation of all the symptoms. The muscles, which are the seat of the trembling, are rigid at the same time, especially the muscles of the neck and shoulders. Pupils dilate and contract naturally. Mouth incompletely closed, and saliva dribbles out of both corners over chin. Articulation embarrassed. Sensibility everywhere normal; muscles contract, though somewhat feebly, by galvanic excitation. Slight dulness over apex of right lung, with diminution of respiratory murmur. Frequent vertigo; occasional cephalalgia; stools natural; urine alkaline, and contains some pus. Questions answered slowly, but pretty clearly. Physiognomy expressive of indifference and apathy.

June 22d to 24th; severe diarrhœa, and involuntary stools, which yielded to opiate injections.

June 25th; but little sleep last night, and delirium; about 10 A. M. had an epileptiform seizure, during which the head was drawn to the right, right eye turned outward and upward, and the left downward and inward. Eyelids and tongue at the same time continually oscillating; while the muscles of the face were rigid. Upper and lower extremities flaccid, offering but little resistance when moved. Complete loss of consciousness, pulse and respiration weak and irregular; duration of fit, eight minutes.

Between 1st and 7th of July, fresh seizures, after which the trembling ceased for half an hour, and then recurred with increased severity. General sensibility diminished from day to day; and the face had a besotted look, like that of typhoid fever. Abdomen swollen, stools involuntary; the patient lay in a sort of sleep, and it was impossible to fix his attention, answering in monosyllables the questions put to him. Strength failed rapidly, pneumonia set in, and death July 11th.

Autopsy.—Tubercular cavities at apex of right lung, and granular hepatization of lower lobe. In the substance of the right optic thalamus, apoplectic cyst of the size of a small bean, the walls of which contained pigment. The *pons varolii* and *the medulla oblongata* manifestly *indurated.* The spinal cord was firm, and *the lateral columns*, principally in *the lumbar region, presented opaque gray striæ.* On a microscopical examination, there was found, in the substance of the pons varolii and of the medulla oblongata, an abnormal production of connective tissue. The opaque striæ in the lateral columns of the cord were due to the presence of connective tissue in process of development.

In the subjoined tables, sixteen cases of disseminated sclerosis are analyzed with reference to age, sex, anatomical characters, disorders of motility, sensibility, special senses, intellect, and the cause of death. They have been carefully selected from the whole number of cases with autopsy now on record, and may be considered as fair types of the several varieties of this lesion of the brain and spinal cord. Case VII. has points of special interest, and the diagnosis offered some difficulties. The paroxysmal lightning-pains in the lower limbs, the tight band around the waist, slight weakness of sight, impossibility of walking with the eyes shut, and the cutaneous anæsthesia, all pointed to locomotor ataxy. On the other hand, the sight-troubles were developed at a late stage, and were insignificant; muscular weakness in the lower limbs was an early, indeed the first symptom; and, although the patient could not walk in the dark or with his eyes closed, still the peculiar, wild thrusting out of the limbs so constant and proper to locomotor ataxy was wanting. There were, moreover, tremor, rigidity, and convulsive movements of the lower extremities, all of which are exceptional in the latter disorder. Finally, the notion of position was retained, and this very rarely happens in the advanced stages of locomotor ataxy. At the autopsy various sections of the spinal cord showed patches of sclerosis, irregularly scattered over all the columns, particularly in the cervical region, but it was on the posterior columns that they were found of greatest extent, thus accounting for some of the embarrassing symptoms, the lancinating pains, etc.

Analysis of Sixteen Cases of Disseminated Sclerosis.

CASES.	AGE.	SEX.	LESIONS.	DISORDERS OF				REMARKS.
				MOTILITY.	SENSIBILITY.	SP'L SENSES.	INTELLECT.	
I. Cruveilhier. Atlas d'Anatomic Pathologique, Liv. xxii., Pl. II., Fig. 4, p. 22.	37	Female.	Indurated spots on the anterior pyramids, right olivary body, and corpora restiformia. The roots of the hypoglossal, glossopharyngeal, and pneumogastric nerves, gray. Patches of induration on anterior face of cord, on the pons, inferior surface of cerebral peduncles, corpus callosum, and fornix.	At æt. 31, weakness felt in left leg; three months after, in right. Later, superior extremities affected; they are feeble and tremulous, but are still capable of use in taking food. Finally, total loss of motion of all the limbs. Articulation embarrassed; deglutition difficult. When spoken to, the muscles of the limbs and trunk are the seat of involuntary movements, which cause the whole body to tremble.	Natural.	Sight feeble.	Perfect.	Lungs tuberculous. Lobular pneumonia. Bronchitis. — Death from pulmonary disease.
II. Cruveilhier. Atlas d'Anatomic Pathologique, Liv. xxxviii., Pl. v., Fig. 1 and 1', pp. 1 and 2.	38	Female.	Gray degeneration of the cord in the form of patches, more or less large, and in much greater number on the posterior than anterior columns of the cord. No alteration of the nerve-roots. Similar patches on several points of the pons varolii. All have a certain depth.	Eighteen months previously, tinglings were felt in the soles of the feet and in the leg, and, almost concurrently, weakness in the lower limbs. Soon after, the arms began to tremble. Fifteen months subsequently, the patient dragged her legs, particularly the left; both constantly gave way under her. She cannot walk without assistance. The left leg more feeble than the right.	Little or no sensation in lower limbs; sensibility very much lessened in the upper. The inferior extremities have at no time been affected with cramps or convulsive movements.	—	—	Death from pleurisy.

III. Friedreich. Reported by Valentiner. Deutsche Klinik, No. 14, 1856.	21	Male.	Patches of sclerosis on the mamillary tubercles, the cerebral peduncles, pons varolii, corpora olivaria, in the substance of the cerebral peduncles, and the medulla oblongata.	Unsteadiness of movements; staggering gait; violent trembling, happening when the patient is spoken to, and which affects the movements of the hands; similar tremor of the head; later, nystagmus.	Occasional attacks of vertigo; pains in the head, particularly in the occipital region. — Sometimes shooting pains in the legs.	—	At the beginning, mental excitement; toward the end, a kind of stupor.	Death from paralysis of the pneumogastric.
IV. Friedreich. Reported by Valentiner. Deutsche Klinik, 1856, No. 14.	20	Female.	Patches of sclerosis on the surface and in the interior of the pons varolii. Induration of the cerebral substance which surrounds the lateral ventricles. Patches of sclerosis on the cord.	Onset at æt. 17, after sudden chill. Weakness of right leg, afterward of left. Trembling of the hands when they are used, afterward participation of the muscles of the eye and tongue. Speech-trouble. Toward the end total paralysis of the lower limbs.	Sensibility of limbs slightly diminished.	—	At the same time, with difficult articulation, a notable diminution of intelligence.	Death from sloughs over sacrum.
V. Zenker. Zeitschrift für rat. Med. Bd. xxiv., Hft. 2 and 3.	30	Female.	Induration of ependyma of both lateral ventricles. Sclerosis of the cornus ammoni, corpora striata, fornix, peduncles of the pineal gland, corpus callosum, tænia semicircularis, pons varolii, cerebral and cerebellar peduncles, and upper part of the spinal cord.	At æt. 26, weakness and trembling of lower limbs; later, of the hands and head. The unsteadiness of gait prevented any employment, although she could still use the hands. At æt. 30, tremor, not apparent when the patient was quiet, nor during sleep, seized the limb when any movement was attempted; on this, the upper and lower extremities, and the head, began to tremble; in the erect position the whole body began to shake. At the terminal period, complete paraplegia. Embarrassment in movements of tongue.	Violent cardialgic spells at the onset and during the course of the disease. Cutaneous sensibility abolished toward the terminal stage. Muscular sense lessened from the beginning.	Slight diminution of sight from the outset.	—	Death from sloughs.

Analysis of Sixteen Cases of Disseminated Sclerosis.

CASES.	AGE.	SEX.	LESIONS.	DISORDERS OF				REMARKS.
				NOTILITY.	SENSIBILITY.	SP'L SENSES.	INTELLECT.	
VI. Vulpian. Union Médicale, June 7, 1866.	51	Female.	Patches of grayish color, variable size, distinct, situated in one district on the right lateral columns of the cord, in others, on the left lateral columns; and, at certain points, on the anterior columns, and at others, on the posterior columns. Same coloring on the medulla oblongata, where it extends to the lateral portions and posterior surface, and reaches as far as the floor of the fourth ventricle. The lower half of the left olivary body has the same hue. The size of the cord is evidently diminished, and in those points where the gray tint is most extended, there is, at the same time, well-marked antero-posterior flattening.	At æt. 24, the left foot was suddenly twisted, and began to drag immediately afterward. Three years later, after a fall, the right leg becomes weak, then the right arm, and, finally, ten years later, the left arm. Twenty years after onset (1865) the condition of the patient was: permanent contraction of all the limbs; spells of spasmodic rigidity coming on very frequently in the muscles of the trunk and limbs; when one of the feet is flexed, and kept in that position, there is immediately severe trembling, which is very difficult to stop, and impossible at times when the right foot is the subject of experiment.	Slight hyperæsthesia.	—	—	Death from bronchitis.
VII. Charcot.	48	Female.	Patches of sclerosis in the walls of the lateral ventricles, in the substance of the corpora striata, on the surface, and in the substance of the pons, and in the posterior and lat-	At æt. 36, diminished power in the lower limbs, which become easily tired; at æt. 42, walking impossible without assistance. Later, rigidity of the inferior extremities, and, at times,	Two years before onset, darting pains in the limbs during the	Sight weak.	Good.	Death from sloughs.

			eral columns of the cord, particularly in the cervical region. A centre of softening, about the size of a walnut, corresponding to the posterior convolution of right parietal lobe.	convulsive tremors in them. Three weeks before death, apoplectiform attacks; the legs became rather flaccid—could not be raised beyond the bed. Four days later, second attack of left facial hemiplegia, followed by paralysis of the left arm.	night, once or twice a month; later, waist-pains. General sensibility somewhat lessened. Loss of notion of position of limbs. Headache, lasting two or three days each month.			
VIII. Vulpian.	53	Female.	Reddish-gray spots on the pons varolii, olivary bodies, particularly the left; on the corpora pyramidalia, especially the right; in the white substance of the centrum ovale. Patches scattered on the different columns of the cord to within two and a half centimetres of its termination, where the cord was entirely sound.	Invasion at æt. 34. After numerous periods of pauses, improvement, and aggravation, walking impossible at æt. 45. The patient could still stand, but when she did so, all the limbs began to tremble. Severe tremor in the feet when they are forcibly flexed, or when an attempt is made to keep them so. Impossible to flex the thigh on the pelvis, the leg on the thigh, the foot on the leg, on part of patient. Although tickling sole is felt, no reflex movement. The arms began to grow weak at æt. 52; no other special symptom in them.	Lancinating pains in the lower extremities, especially the left; sensibility normal on both sides, or perhaps slightly increased.	—	—	Death from sloughs.

Analysis of Sixteen Cases of Disseminated Sclerosis.

CASES.	AGE.	SEX.	LESIONS.	DISORDERS OF				REMARKS.
				MOTILITY.	SENSIBILITY.	SP'L SENSES.	INTELLECT.	
IX. Charcot. Reported by Bourneville, p. 30.	41	Female.	Patches of sclerosis in corpora striata, on external wall of right lateral ventricle, right optic thalami; in pons, and nearly whole extent of cervical enlargement of cord, except posterior columns, and lateral, anterior, and posterior columns of dorsal region.	At æt. 28, fall during pregnancy; two years after, weakness of legs after long walks; two years more, could not walk without help. Amelioration, lasting six months. Progress resumed, and at æt. 35 little use of lower limbs, which were also stiff. At æt. 39 both upper extremities affected, left more than right. An attempt to carry any thing to mouth caused trembling of right side, and slight tremor of head.	Cutaneous sensibility natural. Feeling of heaviness in lower limbs. Darting pains about fifth dorsal vertebra, below right breast, and around waist.	Sight weak, particularly of right eye; diplopia.	Good.	Death from pneumonia and sacral sloughs.
X. Charcot. Reported by Bourneville, p. 44.	36	Female.	Patches of sclerosis in optic thalami, lateral ventricles, pons varolii, cerebellum, peduncles of cerebellum, left olivary body, and cord; the nerve-cells of anterior cornua are shrunken; atrophy and sclerosis of optic nerves.	At æt. 30, after sight-troubles gait began to be unsteady; there was slight hold of the ground. Paraplegia progressive until total; rigidity of lower extremities; upper extremities become weak; tremor on any voluntary movement.	At outset pains about shoulders and in temples—then numbness of feet—seemed to walk on down. Contactile sensibility lost about insteps and lower third of leg. Analgesia of both lower limbs.	First feeling of bar of wood in eyes; a little while after, dimness of sight.	Good.	Death from pneumonia.

XI. J. C. Morris and Weir Mitchell. Transactions of the College of Physicians of Philadelphia, 1868. American Journal of the Medical Sciences, July, 1868.	62	Male.	Patches of sclerosis on anterior and lateral columns of spinal cord.	At æt. 37, weakness in left lower limb, which, in a few years, extends to right; paraplegia progressive until complete; left upper extremity affected, then right. Patient condemned to perfect immobility; unable to feed himself.	At outset sense of weight, with numbness in left leg; also sensation of a bar, an inch or so wide, around left leg; then in right. Sensibility in its several forms scarcely altered.	—	Perfect.	Tubercles in lungs and intestines.
XII. Ludwig Leo. Deutsche Archiv. für Klinik Medicin, 1868, p. 151.	34	Male.	Patches of sclerosis on walls of lateral ventricles, fornix, cerebral peduncles, ependyma of fourth ventricle, and corpus callosum. Optic nerves sclerotic to chiasma; corpora restiformia; calamus scriptorius; pons varolii. Sclerosis of posterior, anterior, and lateral columns of cord.	At æt. 20, headaches and vertigo until æt. 26, when a right hemiplegic attack; improvement; progressive unilateral paresis; at æt. 28, gait so unsteady that he cannot walk without aid. Another attack, left side; paraplegia progressive; gait tottering; when lying down can extend and flex lower limbs, but the motions brisker than natural; exaggerated reflex movements. Upper extremities but little affected; tongue embarrassed; series of apoplectic attacks; paralysis complete; tremor of hands and tongue; muscular contractions.	Cutaneous sensibility in its several forms lessened—not able to localize sensations felt. Pains along spine, in forehead, and in the legs, extending to the toes.	Diplopia at outset; myopia.	Perfect.	Exhaustion.

Analysis of Sixteen Cases of Disseminated Sclerosis.

CASES.	AGE.	SEX.	LESIONS.	DISORDERS OF				REMARKS.
				MOTILITY.	SENSIBILITY.	SP'L SENSES.	INTELLECT.	
XIII. Charcot. Vulpian, Union Médicale, 1866.	46	Female.	Patches of sclerosis on anterior and antero-lateral columns, with lineal streaks on posterior columns in dorsal region.	At æt. 45, weakness of legs; rapid paralysis of all the limbs and of the trunk; sits up with difficulty, and can scarcely support head when raised from pillow; contraction of lower extremities and left arm. Urinary and fecal incontinence.	Three years before motor-troubles, darting pains through thighs and legs. Sensibility preserved in paralyzed limbs.	—	—	Death from exhaustion, four years from onset. Probably previous syphilitic troubles.
XIV. Charcot. Vulpian, Union Médicale, 1866.	43	Female.	Patches of sclerosis on pons varolii, left anterior corpus pyramidalis; very large patch on right lateral column of cervical region of cord. The motor oculi externus nerve altered.	At æt. 38, feebleness of lower limbs; a year later, vertigo, apoplectic attack, followed by right hemiplegia; three years after, a second attack, followed by contraction of flexor muscles of fingers, and forearm of right side; two years later, a third attack; right upper extremity wholly paralyzed and rigid; paraplegia and permanent extension of lower extremities.	Long subject to flying pains and facial neuralgia. After disease is established, pains in left ham and heel; feeling of heaviness and numbness in lower limbs.	Incomplete paralysis of left motor-oculi externus muscle, with internal strabismus and diplopia. Towards end darting pains in head.	—	Gradual sinking after last apoplectic attack, December, 1861; death, February 7, 1862.

XV. Charcot. Ordenstein, p. 78, and Bourneville, p. 129. (Case reported by Ordenstein; autopsy by Bourneville.)	30	Female.	A small patch of sclerosis eleven centimetres above olivary bodies; a larger on cervical enlargement; two, one on right, other on left posterior column; in brachial enlargement, just outside the origin of sensory-nerve roots. A large patch surrounding aqueduct of Sylvius, with processes extending into pons; several of considerable size on the walls of lateral ventricle, which are lost in the substance of the hemispheres; in cerebellum, a small patch on right side, near the gray matter of the surface; another in pons.	Unsteadiness of gait, amounting to titubation; difficulty of speech; tremor on any voluntary motion, particularly in left arm; cannot carry left hand to head or mouth directly; tremor of tongue; muscular force of limbs fair; tendency to fall backward in walking; no permanent rigidity; no spasmodic muscular contractions.	Cutaneous sensibility slightly increased; darting pains in head. Onset with spells of vertigo, and soon after hysterical globus in throat.	Nystagmus, in right optic disk, beginning of atrophy; in left, slight dilatation of veins; sight a little weak, but variable.	Fair; very emotional.	Tubercles in lungs. Death from phthisis.
XVI. Skoda. Reported by Barwinkel. Wien Med. Halle, iii., 13, 1862. Schmidt's Jahrbuch, No. 119, p. 294; Year-Book of New Sydenham Society, 1863, p. 100.	34	Female.	Walls of ventricles, fornix, pons varolii, medulla oblongata, and spinal cord remarkably indurated. Optic nerves hard and flattened. In some opaque, reddish spots of the brain, the nerve-elements are destroyed by new connective tissue; in the pons and cord there was proliferation of connective tissue, and obliteration of the vessels.* Fatty degeneration of muscles.	At æt. 32, feebleness of lower extremities, and soon after tremor of right hand, and then of left—occurring only on voluntary movements. Both progressive. Speech somewhat indistinct.	Began with vertigo, and pains in head and sacrum. Cutaneous sensibility slightly lessened.	Good.	Good.	Death from small-pox.

* In this case there would appear to have been not alone disseminated sclerosis, but also diffuse sclerosis.—M. C.

B. CORTICAL SCLEROSIS OF THE BRAIN AND SPINAL CORD.

The *cortical* or *annular* form of cerebro-spinal sclerosis has recently been described by Dr. Vulpian, and an illustrative case published by him, a brief of which will be sufficient to show the symptoms and nature of the morbid process.

L. B. was admitted into la Salpêtrière, November 7, 1861, with an affection of the spinal cord which began when she was fifty-two years of age, by weakness in the lower extremities, and was apparently neither preceded nor accompanied by lightning-pains nor sight-troubles. This paresis of the lower limbs gradually, but very slowly, increased, and fourteen years after the onset the patient could only walk with crutches, or when helped by some one. Eighteen months later she could scarcely maintain the erect position, even when leaning on a support. Cutaneous sensibility was, when she was examined in 1862 and 1868, very much lessened throughout the lower extremities; but, in 1868, on the dorsum of the feet it had reappeared, and was nearly natural.

There were none of the ordinary symptoms of sclerosis of the posterior columns; there was but little, if any, ataxy of movement; in walking, the left foot was thrown forward in a somewhat exaggerated manner, but there was really no true motorial incoördination. Lying in bed, she could raise one or both legs without any involuntary deviation, and keep them in that position without their trembling. The muscular force of the lower limbs, tested in the horizontal position, appeared good, except perhaps during the last three months.

In 1866, for the first time, lancinating pains were felt in the lower extremities, accompanied by convulsive jerks; but these pains did not seem to have the distinctive lightning (fulgurant) character of locomotor ataxy. Notion of position of the limbs was perfectly good, and, except some weakness of vision, there were no sight-troubles.

Toward the end of 1867, general tremor of the whole body was noticed, whenever the patient attempted to stand up, or tried to walk.

The symptoms collectively were still evidently very different from those of posterior sclerosis of the cord, and it was recognized that the lesion of the cord, whatever it might be, must, in some respects at least, be other than that of locomotor ataxy. This probability was verified by the autopsy. Instead of fascicular sclerosis of the posterior columns, the cortical layer of the white substance was found sclerotic throughout the periphery and in the whole length of the spinal cord. There was also spinal meningitis, particularly well marked on the posterior face of the cord, but easily recognizable on the anterior and lateral faces.

135 Lexington Avenue,
New York, *June*, 1870.

Col. Jno. Moore U.S.A.
with the compliments of the a

NOTES

ON

THE PHYSIOLOGY AND PATHOLOGY

OF THE

NERVOUS SYSTEM,

WITH REFERENCE TO

CLINICAL MEDICINE.

A.—DISSEMINATED SCLEROSIS OF THE BRAIN AND SPINAL CORD.

B.—ANNULAR OR CORTICAL SCLEROSIS OF THE SPINAL CORD.

BY

MEREDITH CLYMER, M. D. UNIV. PENN.,

FELLOW OF THE COLL. PHYS. PHILADELPHIA, ETC., ETC.

[*REPRINTED FROM THE NEW YORK MEDICAL JOURNAL, MAY,* & June, 1870.]

NEW YORK:

D. APPLETON & COMPANY,

90, 92 & 94 GRAND STREET.

1870.

www.ingramcontent.com/pod-product-compliance
Lightning Source LLC
Chambersburg PA
CBHW022013190326
41519CB00010B/1506

* 9 7 8 3 7 4 2 8 3 0 4 9 4 *